# Let's All Keep CHICKENS!

## THE DOWN-TO-EARTH GUIDE TO
Natural Practices for Healthier Birds and a Happier World

### Dalia Monterroso

Storey Publishing

*To Mother and Daddy, who left the
only home they knew so that someday
I could find my way, in peace*

---

The mission of Storey Publishing is to serve our customers by
publishing practical information that encourages
personal independence in harmony with the environment.

Edited by Deborah Burns, Hannah Fries,
    Sarah Guare Slattery, and Lisa H. Hiley
Art direction and book design by
    Alethea Morrison
Text production by Jennifer Jepson Smith
Indexed by Samantha Miller

Cover and interior photography by
    © Rick Dahms
Additional interior photography credits
    appear on page 185
Illustrations by © Zoë Ingram

Storey books are available at special discounts
when purchased in bulk for premiums and sales
promotions as well as for fund-raising or educational use. Special editions or book excerpts
can also be created to specification. For details,
please call 800-827-8673, or send an email to
sales@storey.com.

**Storey Publishing**
210 MASS MoCA Way
North Adams, MA 01247
storey.com

Printed in the United States,
the interior by Versa Press
and the cover by PC
10  9  8  7  6  5  4  3  2  1

Library of Congress Cataloging-in-Publication
    Data on file

# CONTENTS

# WELCOME!

When I was a kid, I never in a million years thought I would grow up to become a backyard chicken educator. Yet here I am, the self-proclaimed President of Chickenlandia. What is Chickenlandia? you ask. Well, the short answer is that it's an educational resource for chicken keepers, named after my chicken yard. It includes my YouTube channel, my podcast, my blog, and my website.

Through these various forms of media, as well as in-person classes and seminars, I share my core philosophy:

✦ Chickens have an innate ability to thrive with very little overhead, and humans have an innate ability to care for them.

✦ We don't need to purchase the most expensive feed or equipment to be good chicken keepers.

✦ If we can learn to raise chickens in a sustainable and inclusive way, we can actually change the world.

As for the full story of Chickenlandia, which shaped my philosophy on chicken keeping and is why I decided to write this how-to book, it actually began so long ago that I wasn't even there. In fact, I wasn't even born yet. You might be surprised to learn that the first glimmer of Chickenlandia began several decades ago in another land, at the foot of a magnificent volcano called Chingo.

Straddling the border of Guatemala and El Salvador lies sleepy Chingo. With its smoky, deep jungle, Volcán Chingo is the breathtaking backdrop of a small and picturesque town called Jerez (pronounced

Volcán Chingo, near Jerez, Guatelama, where my maternal great-grandparents lived

hair-EZ). In the mid-1900s, my maternal great-grandparents were fortunate enough to own land in Jerez, which was at that time a remote village you could get to only by traveling through the mountains on horseback. My great-grandfather Alberto worked their land in Chingo's rolling foothills. Its rich and fertile soil brought forth corn, beans, squash, and delicious tropical fruits. My great-grandmother Maria was in charge of caring for the animals. It is my delight to tell you that she, of course, raised chickens.

In Guatemala, during Maria and Alberto's time, there were no large mills producing massive amounts of chicken feed. There were no prefabricated coops shipped across oceans or industrial farms crowding 10 birds in 4-by-4 cages. Back

then, in Jerez, there were the people and their chickens, caring for each other in a perfect symbiotic relationship. When Maria and her family finished their traditional meals, they threw all that wasn't consumed to the flock. The rest of their chickens' diet consisted of what the chickens could find. Chickens back then pecked and scratched exactly as they do now, continuously searching for insects and foliage. At night, they slept in a tree. It worked.

Let's fast-forward to 2011, when I brought home my first batch of baby chicks and fell swiftly in love. As any good chicken parent would do, I leaned into all the current books and blogs I could find in an effort to ensure their optimal health and happiness. I learned about what chicken feed is best, what to look for in a coop, and how to protect my birds from lurking predators.

My enthusiasm for my flock was so profound that eventually I was asked to teach a class on backyard chicken keeping at our local community college. That experience paved the way to other classes and seminars, my own website and blog, and my persona as the President of Chickenlandia on the "Welcome to Chickenlandia" YouTube channel and other social media. I never planned to become a backyard chicken educator, but my passion for it turned out to be limitless. It still is.

In 2017, I delivered a TEDx Talk at Western Washington University called "I Dream of Chickens." The talk focused on how we can use our relationship with chickens to better understand each other as human beings. When I was doing research for my talk, I learned a great deal about the traditional ways in which chickens have been and are being raised across the globe. I hadn't touched on these practices in my classes or seminars, but I found them fascinating, valid, and important to know about. The most notable thing I realized is how often humans have persisted through severe conditions with the assistance of the incredibly adaptable chicken. Through war, famine, and other hardships, chickens have not only been by our side, but have also played a critical role in helping us to endure it all. It's been a beneficial relationship that is as captivating as it is heartwarming.

To be honest, at the time, the traditional ways of keeping chickens seemed like a whimsical dream to me. Of course, I respected these practices and was curious about them, just as I am curious about all ancestral ways, but they just didn't seem relevant to me or my students. Since I've always referred to my chickens as beloved family pets, I thought the only way to care for them properly was to spend a lot of money. I had accepted the fact that I needed to have a certain level of income to be a good chicken keeper. I didn't understand that I could embrace a more sustainable and inclusive way of keeping chickens while also cherishing them as much as I do.

But then, everything changed.

Sometimes, there are shared experiences engraved so deeply into our timeline that we forever refer to life in terms of before and after that event. The COVID pandemic that hit the United States with full force in March 2020 was one such event. By the time it fully emerged in the United States, the pandemic had already been traveling throughout the world, blanketing the collective consciousness like a tsunamic wave. Here in the United States, we could see the ripples. The rumblings were beneath us. But in our naivety (mine included), we had no idea of the impact it would have.

I spent a lot of time with my chickens that first month, breathing, crying, and being really afraid.

Throughout social media, worried questions echoed across screens. How will we get through this? How will this affect our children? How can we stop feeling so hopeless? As the anguish dragged on and our disparities reached a tipping point, this collective anxiety turned to anger. I witnessed friends turn against friends, family against family, and a divide grow so deep that I began to wonder if it would ever heal. I'm only one person, I thought. How can I find the light at the end of this turmoil when all I know about is caring for pet chickens?

From that place of desperation, the memory of Alberto and Maria arose in me. I remembered that so long ago, as they farmed their little piece of land, a

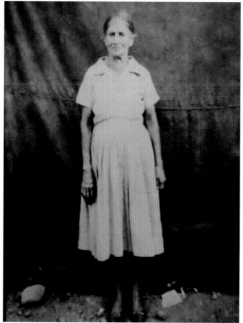

My great-grandmother, Maria Teresa Moran Cotto (at top), and my great-grandfather, Alberto Cotto Aguilar, raised chickens in Jerez, Guatemala, in the mid-1900s.

democratically elected president was overthrown, and their nation crumbled into civil war. Crime took over the streets, malnutrition stole the lives of both the young and old, and there was a horrific genocide of the Indigenous Guatemalan people. Their country experienced anguish that I can't even begin to imagine. It was so great that my mother and father, who lived in Guatemala City, reluctantly left the only home they knew and moved to the United States sight unseen. Yet Maria and Alberto remained steadfast on their land. Is there something about their traditional ways that could help us through our current chaos?

I'd been working on this book for quite some time, but after the pandemic hit, I decided to go back and change many things about it. How I understand chicken keeping now is so different from how I understood it only a few years ago. I've learned that in unstable times, a communal shift toward sustainability is more necessary than ever. This means that self-sufficient practices must not be limited to those who reside on big plots of land or enjoy a certain income level.

*To keep chickens is to participate in something that is part of our collective history.*

From the suburbs to the big cities, I want to see people of all tax brackets, races, and cultures experiencing the joys and benefits of chicken keeping. By sharing a hybrid of modern and traditional practices, I hope this book can increase the accessibility and sustainability of chicken keeping in a more impactful way.

In the following pages, you will notice that I've chosen not to label the method of chicken keeping I teach. My reasoning is this: My aim is not only to keep things simple but also to serve as a reminder that the way many of our ancestors existed with the land and their animals was not a method they read about and decided to use; it was simply *who they were*. Many of their farming practices are now being "discovered" as worthy, or even rebranded and introduced as new ideas. I cannot claim that the ideas I share are a new discovery, and I don't wish to call what I teach anything other than chicken keeping, because that's what it is. I am not married to any one idea or system. I simply seek to reconnect you with something that was always there, within you.

To me, chickens aren't just livestock. They are an heirloom passed down by both my ancestors and yours. To keep chickens is to participate in something that is part of our collective history; it's a common link that not only reflects how we as humans have chosen to live but also how we've managed to survive. That's why the story of Chickenlandia isn't just mine; it belongs to all of us. As good tales go, it continues to change and collect meaning. In the time of Maria and Alberto, Chickenlandia began with a wish for peace in an uncertain world. What wish will chicken keeping hold for you?

Welcome to Chickenlandia. Welcome to where you've always belonged. Let's get started.

# THE CHICKEN AND HUMAN REUNION

When the backyard chicken bug first bit me, I was eager to pick out breeds, research coop design, and plan for my first flock of baby chicks. Thankfully, I had an experienced mentor who brought my attention to some preliminary questions that were important for me to answer before I started planning in earnest. To save you from any unpleasant surprises, I suggest you explore the poultry laws in your area and examine solutions to any pushback you might face from your neighbors. It's also essential that we go over the unfortunate dilemma of backyard roosters, and why you should, shouldn't, or simply can't keep them.

## Looking Back to Go Forward

Did you grow up around chickens? I didn't. Like many of you, I grew up separated from my food. Breakfast, lunch, and dinner were reflected to me not in the barnyard, but on a small, bulbous screen with memorable jingles and characters meant to catch a child's eye. My mother, an incredible cook, whipped up fragrant Guatemalan dishes between her shifts as a data-entry employee. But I didn't want those traditional meals. I wanted what was in the commercials. I used to sit alone on my porch and wonder why we couldn't be like my friends' families, the "otherness" of my existence palpable. Among the neighborhood ambience was a rooster's crow, barely audible in the distance. It was like

a call to something I once knew about but could no longer understand.

I was born in the seventies and lived my entire childhood and teenage life in a small suburb of Dallas, Texas, called Duncanville. Even though we were considered poor, I have fond memories of the humble neighborhood I lived in. My best friend and her family lived just a few houses down from ours, and there was a beautiful little park within walking distance called South Side Park. I wasn't allowed to go to South Side without an adult, but my best friend's family had more lenient rules, so she and I would go there from her house almost daily (sorry, Mom).

There, among the oak and redbud trees, we would announce visions of our future as children do. "I'm going to change

Four-year-old me in front of my childhood home

the world!" I used to proclaim. There were deafening cicadas, horny toads, tiny snakes, and grasses that went up to our waists. When we were near the pond we would tread lightly for fear of water moccasins, the legend of the one kid who was chased and bitten always fresh on our minds. We never did see one.

At the farthest end of the park, past a small strip of forest, there was what my family referred to as a "rich" neighborhood called Swan Ridge. Sometimes we would get fairly close to that section of the park's edge, but it always felt like that was the one line we should never cross. It's funny because some of my closest friends over the years lived in that so-called fancy neighborhood, but I never walked to their houses, and they never walked to mine. It was clear there was more than a park between us. Riding to each other's houses in a car instilled that invisible partition.

On the closest edge of the park, on the other side of a baseball diamond, there was an unkempt alleyway lined with the mismatched fencing of other people's backyards. One plot was particularly overgrown; the dry grass was probably 3 feet tall. You could see various discarded objects, like a rusted lawn mower, a child's faded ride-on toy, and some kind of makeshift wooden box. What I remember the most about this particular home is the sound that came from it. You couldn't see the feathers through the foliage, but you could hear the hens singing and the rooster's proud crow.

I don't know what the laws concerning chicken keeping were in my neighborhood or Duncanville in general. But I do know that no one in our neighborhood was about to complain about a rooster crowing. I also know that if that same rooster had existed in Swan Ridge, he would have been evicted in short order. Why is that? At what point did the idea of raising chickens symbolize poverty and the so-called lower class? When I was just getting started in my chicken-keeping journey and telling everyone who would listen about it, I literally had someone laugh and say, "Dalia, this isn't *Guatemala*!" No, it isn't. But maybe it's time we took some lessons from those of lesser means.

## Can You Even "Chicken"?

At a critical point in US history, those who lived in urban areas were separated from their food. This separation started in the 1920s, when public health agencies began outlawing livestock within city limits. Their reasoning was valid; there were justifiable concerns regarding public health and safety. But this is also when large-scale factory farming entered the scene, with the chicken as its first victim. Why were factory farms the solution for feeding us, when chickens are versatile and can be kept responsibly and considerately in a backyard or community setting? I can't help but think that if we had focused our laws on solutions that didn't line the pockets of the few, we would be in a much

There's no reason chickens can't once again become part of the urban landscape, as they are in this photo from the early 1900s.

better position now. But there's still time to change things, and you can be part of the solution.

So here's my first question to you, and it's *super* important: *Are you allowed to have chickens where you live?*

If the answer is yes, that's awesome! If the answer is yes but you can't have a rooster, that's not quite as awesome but it's better than nothing. If the answer is no, we definitely have some work to do.

To find out what the laws are in your area, you will need to check your municipal codes. You can find these on your local government's website, where there should be a section regarding animals

and livestock. Make sure you read very carefully, as there will be lots of legalese that can get confusing. An alternative (yet sometimes less reliable) option is to call the main office number to find out the rules. If keeping chickens is illegal in your area, here are some easy steps you can do to work on changing things.

1. If you know anyone in your local government office, talk to them first. Having an ally and a direct line of communication can be very helpful when figuring out what steps to take specific to your area. It could also tip the scales in your favor when it counts.

2. Find out what the laws are in neighboring counties. It will be good to have examples of how chicken keeping can be workable and beneficial to your city or town.

3. Put feelers out on social media for others in your area who would also like to see the law changed, and ask them to join the effort.

4. Gather information and favorable testimony regarding the main concerns about chickens, which are usually noise, smell, rodents, and property value.

5. Petition the city council for an ordinance change and be prepared to compromise. You may have to start small with the knowledge that the steps you take now can change the future ordinances of your town and the nation.

Chickens can be kept responsibly and considerately in a backyard or community setting.

## COMMON ARGUMENTS AGAINST KEEPING CHICKENS

As you set out to change your town's laws, it's important to know the arguments you will come up against from people who want to keep chickens out of their neighborhoods. As a compassionate President of Chickenlandia, it's important for me to try to understand where other people are coming from and put my educator's hat on rather than simply invalidating their concerns. I urge you to do the same, although I understand how frustrating these baseless arguments can be. So let's break down the top three.

### Chickens Attract Rodents

Contrary to what many might think, rodents—especially mice and rats—are literally everywhere. I live next to a forested area, and every night I can almost guarantee their presence to some degree. Even in large cities, the rat has effortlessly inherited every nook and cranny, an expert at cleaning up after their most forthcoming host: the human. In short, everywhere there are people there are rodents, and beyond that, there are still rodents!

Though generally underappreciated, the rat is essential to the delicate balance of our ecosystem. Who else is going to clean up all that leftover pizza in the New York City subway? In the city and country, rats keep insect populations in check, clean up decaying material, and contribute to the diets of other animals in the natural world. The fact that we don't see their value doesn't change their contribution to our lives. The truth is, these animals really don't become a problem until we create the right conditions for an infestation.

There is no denying that mice and rats love chicken feed, kitchen scraps, and even chicken droppings, but they also love bird feeders and compost bins, though the latter two are rarely outlawed or acknowledged as a problem. The good news is that mindful husbandry, strategic feeding, and securing your flock's feed can decrease a rodent's temptation to start a family in your backyard. Also, you can administer kitchen scraps in a way that rodents can't access them, thus deterring the rodents. In short, if we take the right steps, our neighbors can rest assured that these "pests" will remain out of sight and mind.

### Chickens Stink

A properly kept chicken coop and run should not smell, period. Yes, chicken poop is stinky when fresh and when you're up close and personal. But if the chicken

Chicken droppings and bedding can be transformed into rich fertilizer.

keeper maintains good practices, the smell, unlike the odor of dog poop, isn't lingering or far-reaching. Also, unlike dog poop, chicken poop can be easily transformed into a rich fertilizer, which you can then use to grow delicious vegetables. This is not to knock dogs! I have four rescue dogs (yes, you read that right: four) that I love dearly. I just want to bring attention to the arbitrariness of a portion of society's negative feelings toward chickens. My first chicken coop was in the same place for over 10 years, and I never had a complaint regarding smell or anything else, for that matter. With the right rules and oversight, everyone can coexist peacefully with chickens.

## Chickens Are Noisy

Well, yes, they can be. Even in the absence of roosters, hens can start singing their "egg song" rather early and in fact be quite loud regarding their accomplishments. But let's talk about chicken sounds in comparison with other neighborhood sounds that are generally accepted. A barking dog can reach upwards of 100 decibels (dB), while a lawn mower can achieve around 90 dB. A motorcycle engine is around 96 dB, and a child screaming while playing can be up to 120 dB! The hen song by contrast is around 70 dB. I'm not even mentioning the usual city sounds like sirens, horns, and eccentric people shouting. Hens don't sing all day long. So why, exactly, is their specific sound an issue?

If you find out that you can have chickens, make sure to ask if there are any restrictions in number, and what the rules are regarding roosters. Find out if there are any guidelines you must follow as far as coop and run placement, size, and appearance. Find out what the penalties

Chatty hens are a reality, but their noises can enhance a dynamic neighborhood ambience.

could be should you fail to meet the requirements, and what might happen if a neighbor complains about your flock. Get all the information you can before you get those first baby chicks so that you don't find yourself in a difficult situation later.

## Cock-a-doodle—DON'T?

It's quite likely that in your research you've discovered that roosters aren't allowed in your town or city. You may or may not agree with this, but I'm going to go out on a limb and express why I think we absolutely need roosters to crow in all neighborhoods across the nation. I hope you'll bear with me and keep an open mind (and open ears).

Roosters get an exceptionally bad rap among those who don't live in the country. I promise I'm not about to speak poorly of city-dwellers. I mean, I'm from the suburbs and have lived in many cities, but I know that urban feelings about roosters are notoriously misguided. When people talk about roosters crowing in *their* neighborhood, you would think it was the screech of some malevolent alien species coming to devour their sleep. Yet the call of a full-grown, standard-size rooster lands at around 90 dB. That's about the same noise level or less than you'd get from a barking dog. So tell me again how this specific sound is somehow the loudest thing ever?

The undesirability of roosters creates an unfortunate dilemma for hatcheries, the companies that supply baby chicks to farm stores, farms, and individuals. Hatching eggs have about a 50 percent male hatch rate. Have you ever wondered what happens to the male chicks that are in far less demand? It's something that I don't like to think about. Honestly, I don't even discuss it much in my teachings because there aren't a lot of options available for those who can't have roosters but also don't feel comfortable eating unwanted birds. So I'm stuck, and so are the hatcheries, which I'm sure would rather have more options as well. I've also comforted many families who were forced to give away their pet roosters due to complaining neighbors and/or city laws. I would love to see that change.

If you aren't allowed to have chickens where you live and you're working on changing the laws, you will likely need to make some compromises. This might mean crafting regulations that exclude roosters, and that's okay for now. Hopefully, in the near future, I can update this chapter to celebrate the inclusion of hens *and* roosters in every neighborhood of the country.

I've had to make a lot of changes in my life as of late, most of which I never thought I would have to make. Change is difficult, but it is more necessary than ever in our current times. When I think about the inaccessibility of chickens in the city and suburbs, I see more than just an inconvenience. I see a system working toward keeping us separated from something essential, not only to us as people but also to the earth we live on. We've been tricked into thinking that keeping chickens among us is a negative thing, and it's time for us to shift our viewpoint even if it means being woken up early by the occasional cock-a-doodle-doo.

The fact that the human/chicken separation affects marginalized communities the most is not lost on me. But we can still make it right. In my mind, a rooster's crow is the sound of old wisdom returning and an old wound healing. It's the sound of a more sustainable future for our children. Let it ring in a new beginning for all of us.

A rooster's crow can be the sound of a new beginning.

# PLANNING FOR CHICKENS
## with All Living Things in Mind

The first questions I'm usually asked when someone is preparing for their first flock are probably the same ones you're wondering about. What breeds are the quietest? What breeds are the tamest? Which breeds lay the most eggs? These are all great questions, and I'll answer them in this chapter. But now that I'm several years into this chicken thing, I've realized that when someone is in the planning phase, it's of the utmost importance that they understand how all their decisions might affect the natural world. While I'm not completely against the use of synthetic medications or pesticides when absolutely necessary, my goal is to reconnect people with practices that are more natural, sustainable, and in line with what our ancestors did.

## Your Yard Is an Ecosystem

If you were to walk outside and into your yard right now, no matter how large or small it is or the amount of concrete or plant life it contains, you would immediately be immersed in a delicate ecosystem. Sure, there are human-made items there—perhaps a set of lawn chairs, a garden hose, or maybe even a gnome or two. But there is also a complex network of biotic (living) and natural abiotic (nonliving) things, each existing to create balance in your mini-world. And it doesn't stop with your yard. If you were to take a peek over your neighbor's fence (I don't recommend actually doing this), you would find another ecosystem tightly connected with yours, and so on and so forth. In fact, our whole planet is one big ecosystem.

As you embark on planning your flock, I want to first introduce you to the idea that once you bring chickens home, they become part of the ecosystem in your backyard and beyond. There are both natural and human-made ecosystems, the former being things like oceans, prairies, and forests; the latter, things such as cities, gardens, and farms. The problem with many human-made ecosystems today is that they don't make much of an effort to create a working relationship with nature, and thus they tend to disrupt the natural world rather than cooperate with it.

You will inevitably alter the natural patterns of your area through your chicken-keeping practice since you are introducing non-native animals and building human-made structures in which to house them. But it is my hope that ultimately your practices will cooperate with and respect the environment all of us live in, just as many human practices have done throughout history and some still do throughout the world. I think this will offer the most benefit for you, your chickens, and the earth.

The ecosystem of a chicken yard includes essential members of the microworld such as bacteria, fungi, and protozoa.

Some of the members of your backyard ecosystem that could be affected by your chicken-keeping practices are so tiny that you can't even see them with the naked eye. Their size, however, does not reflect the magnitude of their importance. All over the earth, in our chickens' bodies, and even in our own bodies, there is an invisible universe of microorganisms such as bacteria, fungi, and protozoa, and we have only scraped the surface of their significance.

As you become more familiar with mainstream poultry practices, you will inevitably hear about pesticides, antibiotics, and other synthetic medications that are commonly recommended and used. What is not widely discussed is how these substances can disturb the fragile microbiomes of our environment and create an imbalance in our local ecosystems and beyond. We can see this problem on a large scale when observing the effects of modern agribusiness's overuse of antibiotics, which has resulted in the proliferation of antibiotic-resistant bacteria.

## Six Rules for Mindful Chicken Keeping

The following is a list of commonsense rules that take into account the benefits of all living things. Think of these as a foundation for all the practices we will dive into. While it's best to keep these considerations in mind when you get your first chicks, it's never too late to evolve the way you keep chickens.

1. **Listen to your heart.** When in doubt, follow your intuition. Remember that humans have been keeping chickens for thousands of years and this ancient knowledge exists within you.

2. **Chickens belong outside in nature.** Let your flock have contact with the natural world from chick to old age. That's where they have existed for millennia, and it's where they are supposed to be today.

3. **Focus on balance.** Health is attained in a balanced state. Keeping your flock's emotions, environment, and nutrition balanced is the key to avoiding issues.

4. **Choose natural options whenever possible.** Synthetic medicines and pesticides have a time and place, but they shouldn't be your first option unless they're your only option. Natural methods and modalities are generally less disruptive to life and the environment.

5. **Choose practices that are compassionate.** Making compassionate choices for your flock extends to your fellow humans and all the creatures of the earth.

6. **Live in harmony with wildlife, big and small.** The ecosystem in your chicken yard is complex and intelligent. Seek to limit disturbing it.

## Terms and Considerations for Beginners

Now that you've made a commitment regarding how you want to keep chickens, there are some crucial things you need to consider before you take the plunge into chicken-hood. None of this is meant to discourage you, but as someone who rescues chickens and volunteers at my local Humane Society, I have known too many abandoned hens and roosters. I'm also painfully aware of the numerous roosters that get dumped in the woods in the spring and summer when new (and experienced) chicken keepers realize they have a crower on their hands. These things are tough for me to witness, but as I've stated before, my priority is not to judge, but rather to *educate*. I hope that in sharing the following information, you can also help to spread awareness.

Chickens born in an industrial hatchery or on a breeder's farm are sometimes sexed soon after hatching. This means that someone examined them to determine whether they are male or female and then separated them accordingly. Male chickens who haven't reached maturity are called cockerels, while females who haven't started laying are called pullets.

When you order chicks through a hatchery or purchase them through a breeder or at the farm store, you will notice that they are most often divided into two categories: pullet and straight run. If the chick bin has no distinctive label or says *straight run*, you may want to run straight out of that store! I say this because *straight run* means the chicks have not been sexed, so you would have a very good chance of getting a rooster or roosters. If you are looking for bantam chickens (more on those on page 30), know that most are sold as straight run because their small size makes it difficult to determine their sex at hatch.

Bantams are usually sold as straight run due to the difficulty in sexing small chicken breeds.

While I do think roosters should be allowed in every suburb and every city, this doesn't change the fact that a flock needs more hens than roosters. My general advice is to have no more than one rooster per at least eight hens. Now, there will always be exceptions; some roosters are more docile than others and can coexist with fewer hens without harming them or their fellow roosters. Also, some breeders find that they have no issues with keeping certain breeds in pairs. But for the most part, even if you are allowed roosters where you live, you will need to have a plan if you end up with the wrong ratio.

Finding a new home for a rooster is usually not too difficult thanks to social media and various online classifieds, but if you don't want your rooster to be eaten, you need to state that upfront. And you should be aware that in the United States and many other places, cockfighting is unfortunately still a thing. Some people will take in unwanted roosters for the purpose of bait or fighting, so don't be afraid to ask questions and really vet whoever shows an interest in adopting your bird or birds. You won't regret it.

Hens have a laying life and an actual life span, the former usually being substantially shorter than the latter. Some chickens can live 10, even 15 years with good care, though breeds bred for laying usually don't make it that far. Their best laying will occur in the first year of life, will still be great the second year, then drop

Hens lay most prolifically in the first couple of years of their lives.

off each consecutive year. Before you get chickens, it's important to ask yourself what you will do with aging hens that no longer lay. In Chickenlandia, my flock lives out its days without any conditions, but I'm not running an egg business or following a super-tight budget. I completely understand if keeping aging hens wouldn't work for you. The only thing I ask is for you to have a humane plan once your hens slow down or stop laying.

While some breeds are considered generally quiet, every chicken has its own personality. I'm sure you're familiar with the quintessential hen sound that goes something like *Bawk, Bawk, BagAWWWK!* This is known among chicken enthusiasts and farmers as the egg song, and it's sung by many hens (sometimes the whole flock) after an egg has been laid. A hen may sing the egg song loud and proud each and every time she delivers breakfast for you, regardless of her breed. It's for this reason that I believe in informing your neighbors of your plan to get chickens *before* you get them, especially if your flock may include a rooster. If they voice their displeasure, you will know where they stand and can prepare accordingly. This is where knowing your local ordinances really matters. Know your rights and, whatever you do, always promise eggs. You'd be surprised how much the taste of fresh eggs can quell a cranky neighbor!

Now that we've gotten that out of the way, let's talk about you.

## What Kind of "Chicken Person" Are You?

The type of chickens you should adopt into your life has more to do with you as an individual than it does with the different breeds. Ask yourself: Why are you getting chickens? Do you want a thriving, year-round egg-selling business? Do you want to keep your family and perhaps a few neighbors rich with eggs? Perhaps you don't even eat eggs, but just can't resist those fluffy butts? I understand all of these scenarios, and I think the variety of chickens you go for should reflect the answers to these questions.

## STANDARDS: FOR THOSE WHO WANT LOTS OF EGGS

There are two sizes of chickens: standard (also called large fowl) and bantam. Standard chickens are the ones you've probably seen pecking and scratching in the background of your favorite films and TV shows. They can weigh anywhere from 4 to even 12 pounds and are commonly divided by hatcheries, breeders, and hobbyists into two unofficial categories: heavy breeds (above approximately 6 pounds) and light breeds (below approximately 6 pounds).

Heavy breeds can usually do better in colder climates than light breeds but can be more susceptible to heat exhaustion, with some exceptions. Their generally smaller combs are not as susceptible to frostbite but don't release heat as efficiently either. Heavy breeds are great layers and typically more broody, meaning they have more of an instinct to sit on and hatch out baby chicks (we'll talk about this in more detail later). They also often have less foraging ability than light breeds.

Discernibly lighter breeds, whose combs are comparably larger, will often fare better in warmer climates than extreme cold. Lighter breeds tend to be more flighty than heavy breeds yet better at foraging, while heavier breeds seem to prefer to stick closer to the ground (and their food bowl). Light breeds can also be phenomenal egg layers and are generally less broody than heavy breeds.

There are both heavy and light standard breeds that can do well in either hot or cold

**Heavy breeds** are bigger, have smaller combs, and are usually cold-hardy.

**Light breeds** are smaller, generally have larger combs, and are often heat-hardy.

climates if some provisions are taken when temperatures become extreme, which we'll cover in Chapter 11. Some standard chickens are considered dual purpose, which means they can be kept both for laying and meat, though I won't be covering meat birds in this book. Chickens bred for optimal laying fall under the standard classification.

Giving special consideration to the factors above, along with *general* quietness and optimal egg-laying in mind, I've put together the following list of breeds that I think are the best for new chicken keepers who want to keep chickens as a source of eggs and companionship.

**Barred Rocks.** These chickens are often seen in movies and television, and for this reason I consider the breed to be the quintessential chicken. Developed in New England, this heritage breed has earned its popularity with its prolific egg-laying (large and brown) and its resilience. The birds are quirky and fun and don't often go broody.

Barred Plymouth Rocks are 7 to 9 pounds, cold- and heat-hardy, with a single comb and yellow legs.

**Buff Orpingtons.** Developed in the UK, the Buff Orpington is a very good layer of large, light brown eggs. These birds are cold-hardy but will need a mindful eye in the summer to ensure they don't get overheated. Known for its fluffy butt feathers, this is one standard breed that does have the tendency to go broody and has an innate ability to be an excellent mother. Buff Orpingtons are normally very docile, sweet, and easily tamed.

**Black Australorps.** These chickens were developed in Australia from the Black Orpington, making them very similar in temperament to their cousin, the Buff Orpington. In the sunlight, their striking black feathers reveal a purplish-green sheen. They are champion layers of large, light brown eggs. They are less broody than the Buff Orpington but do make excellent mothers if given the opportunity. Friendly, calm, and generally quiet, Black Australorps should be easy to tame, especially if handled as chicks.

Buff Orpingtons are 7 to 10 pounds, cold- and heat-hardy, with a single comb and yellow legs.

Black Australorps are 7 to 8 pounds, cold- and heat-hardy, with a single comb and slate legs.

**Easter Eggers.** Known for their colorful eggs, Easter Eggers are not a true breed and as such do not conform to any breed standards. But that doesn't make them any less fun! Developed from the Araucana and the Ameraucana, they will lay one color egg for life, either green, blue, pink, or brown. Many have cute little beards, giving them a delightful, owl-like quality. Easter Eggers are often sold as "Americaunas" or "Americanas" by hatcheries and farm stores. They are good layers and usually very friendly.

Easter Eggers are 5 to 6 pounds, cold- and heat-hardy, with a pea comb and slate legs.

## BANTAMS: FOR BEAUTY AND FUN (AND SOME EGGS)

Domestic bantam chickens are said to have originated in Bantam Village on the island of Java. In modern times, they are considered any chicken of a smaller stature compared to standards. They weigh anywhere from a few ounces to around 3 pounds and can be miniature versions of standard breeds (pro tip: it's adorable to have both variations in your flock) or "true" bantams, which means there is no standard counterpart.

I've been asked numerous times if bantam eggs are okay to eat or if bantam chickens even lay at all. The answer is yes and yes! Most people find their eggs adorable, since they are perfectly "kid-size" and fit flawlessly on an English muffin. I know one farmer whose bantam eggs sell out at every farmers market because customers find them so adorable.

Though bantam eggs are delicious, bantam chickens, in general, are bred more for beauty, personality, and broodiness than for laying frequency. A chicken is considered broody when she has a strong instinct to sit on, hatch out, and raise baby chicks (in other words, brood babies). While broodiness might be a desirable trait to some, if you want chickens solely for eggs you may not consider this a useful quality, since sitting on eggs means they will not lay during that time. Recently, some hatcheries have recognized the attractiveness of a compact layer in urban

STANDARD EGG

BANTAM EGG

All eggs have some natural variation in size, but generally you can substitute two or three bantam eggs for one standard egg.

areas and so have begun breeding for that trait, but no bantam will lay as well as its larger cousins.

I have mostly bantam breeds in my flock, and many people ask me why I keep so many. To put it simply, bantams are a ton of fun in a small package. My family loves them for their silly personalities, cuddliness, and beauty. Some breeds of bantams, like Serama, aren't ideal for beginners due to their flightiness and special needs. Here are some beginner bantam breeds suitable for those just starting with chickens.

**Silkies.** These are perhaps the most popular breed of bantam chicken in the United States, although their classification as bantams differs according to who you talk to and where you are in the world. Often referred to as the "dogs" of the chicken world, Silkies are furry, friendly, docile, and sweet. Thought to have originated in China, Silkies are a very old

Silkies are 2 to 3 pounds, heat-hardy, with a walnut comb and black legs (with five toes on each!).

breed but exactly how old is unknown. Throughout history, they have been prized for their unique black skin, bones, and even flesh. They lay small, cream-colored eggs, are very broody, and make excellent mothers.

**Bantam Cochins.** Sometimes referred to as Pekins in the UK, though there are some who believe Pekins to be a different breed altogether, Bantam Cochins also hail from China and are beloved for their sweet personalities, fluffy butts, and feathered feet. They lay light brown eggs that are surprisingly large for the breed's size, and they definitely love being broody and raising baby chicks. The non-frizzled variety of Bantam Cochins is very cold-hardy, but it doesn't do as well in the heat.

**Bantam Wyandottes.** These birds were developed in the United States, are friendly and docile, and lay a decent number of relatively good-size brown eggs. They are stout little chickens and can be a tad chatty, but they are so funny and cute that I doubt anyone would mind. Bantam Wyandottes are very cold-hardy but don't do well in extreme heat. They will go broody, but not as often as Silkies or Cochins.

Bantam Cochins are 1 to 2 pounds, cold-hardy, with a single comb and yellow legs with feathered feet.

Bantam Wyandottes are 1½ to 2 pounds, cold-hardy, with a rose comb and yellow legs.

## MIXED FLOCK: FOR CHICKS, FUN, AND EGGS

Of course, you don't have to choose between standards and bantams. If you dream of a mother hen with fuzzy babies in tow or you just want a little extra cuteness in your chicken yard, include some bantam breeds in your standard flock. If you were to visit a larger-scale egg farm, you might see some Silkies or Bantam Cochins running around among the working layers. That farmer knows that keeping a few bantams will make it easier to hatch new layer hens each spring, since broody bantams will gladly mother chicks of a larger size. Why add the extra workload of incubating and raising baby chicks when you can have a hen do it for you? I'm sure the farmer appreciates the beauty and joy that bantams bring to a flock as well.

You don't have to choose between standards and bantams. Why not have both?

Starting out with a flock of no fewer than four chickens can reduce pecking order issues.

## How Many Chickens Should You Get?

When I was preparing for my first flock, I told The First Man (my husband) that I would only be getting "three or four" chickens. By the time I got to the farm store to pick them up, I think I had settled on 6 chicks. Then I came home with 10.

I laugh at this memory now, because I understand the full meaning of "chicken math." So many times, I have witnessed someone's dream of 3 or 4 chickens turn to 8, which then turns to 12, which eventually turns to 16. While it's funny to think about, starting out with too many chickens can be stressful and end up creating a discouraging situation for beginners. I know this

### Beginner Flock Size

✤ **Number of chickens:** minimum four chickens (one layer per family member)

✤ **Coop space:** 2 to 4 square feet per standard chicken

✤ **Run space:** minimum 10 square feet per standard chicken

from experience. So let's talk about how many chickens you should start out with to make the beginning of your chicken journey an easy one.

While the right number of chickens for you depends on a few things, the minimum I usually suggest is four. Chickens are flock animals, meaning they need the dynamics of a multi-member flock to feel happy and secure. If you start out with fewer than four chickens and you end up losing one or two, you could have problems when you attempt to replace them if you are down to one chicken.

This is due to strong flock instincts and a pecking order, which refers to a system of hierarchy among flock members, that can make introducing new chickens to a single bird very challenging. It can certainly be done, and we'll discuss it more in a later chapter, but it's a multistep process that can cause stress on new chicken keepers and their birds. If you start out with four flock members and you lose one or even two, integrating new chickens will be easier on you and your flock.

If you want chickens strictly for eggs and you're starting out with a small flock solely for your family, I recommend beginning with one chicken per family member while observing the minimum of four chickens. If your family eats lots of eggs, or if you want to give away or sell some to your neighbors, you will want to go higher than that. But it's in your best interest to start small and plan for growth, especially when you calculate the power of chicken math!

You'll also have to consider the size of the area you have to offer them. A good rule of thumb for standard-size chickens is to give each one 4 square feet of space inside the coop and at least 10 square feet of space in the run. You can get away with 2 square feet per standard chicken inside the coop *if* the chickens have additional covered space that is sheltered from the elements and they are basically limited to the coop only at night. Of course, you can get away with having many more bantams than standards in the same size space. Depending on their size, you could probably fit two or three bantams in the place of one standard-size hen. That's another reason why I love bantams so much!

## Change Your Mind, Save Your Change

Chicken keeping has been one of the most beneficial agreements between humans and animals that we have ever made. For thousands of years, chicken keeping was a completely sustainable practice, with humans feeding their flocks scraps that would otherwise go to waste. Chickens, in return, provided food, and, once farming began, fertilizer. While sustainable chicken keeping still exists today, it's disappearing from the mainstream even in poor areas where there is a movement to "educate" people about the advantages of so-called modern ways. These ways require the purchase of feed, medications, expensive housing, and equipment. What happened? How did chicken keeping go from a sustainable practice to something that actually uses resources and has become out of reach for so many?

Certainly, industrialization and the advent of factory farms are the main culprits in the transformation of chicken keeping from something that everyone can afford to something that requires a certain income level to do "appropriately." As industrial farms took hold, their standards and practices became the lens through which we received information on how to care for our chickens. Even today, most scientific research on chickens does not occur in backyard flocks but rather under factory conditions. Of course, those two environments could not be more different—in one, chickens enjoy sunshine, fresh air, and a good quality of life, and in the other, they enjoy none of these things. I dare say it's not unreasonable to question whether we should apply science that has been conducted in factory conditions to birds that delight in a backyard setting. Regrettably, in many cases, that science is all we have. While I'm certainly

Chickens and people live side by side all over the world, as shown by this flock in sub-Saharan Africa.

not suggesting we deny solid scientific research, unfortunately, much of it doesn't apply to backyard chickens.

Here are some of the recommendations that have come from the research conducted at factory farms: Don't let your chickens eat anything other than their processed feed, don't allow them to have contact with the microbes in the soil, give them preventive medicines every month to avoid disease, treat them with synthetic pesticides for parasitic infestations and infections, and cull them at the first sign of illness or injury.

All these rules are sound when applied to chickens in battery cages, barely hanging on to life in the first place. But for backyard chickens, they not only do more harm than good but also raise the price of chicken keeping considerably. Your feed costs alone would be colossal under this model, not to mention the long-term environmental price that is paid.

I'm not against feeding chicken feed (and in fact, I recommend it), nor do I think that you should avoid all medicines for your chickens. But I think we can make smarter decisions. Let's say you supplemented your chickens' diet with some healthy kitchen scraps, plus sprouts and fodder you grew yourself. Perhaps you also ferment their feed (we'll discuss this in Chapter 7), which can potentially double its volume and increase its nutritional value, resulting in less dependence on big manufacturing. Do you see how adopting some age-old practices can not only decrease the bottom line but also make for healthier, happier chickens?

In the following chapters, I'm going to talk about coops, runs, nutrition, supplements, and all the other things chickens need to live a happy, healthy life. I'll be sharing ideas for products you can buy as well as cost-effective alternatives. Remember, you don't need a lot of money or resources to raise chickens, and you likely know more than you think you do. As I look back at a history of chickens kept by those rich and poor, I can see that to move forward, we need only to "unlearn" some mainstream ideas.

# YOUR FLOCK CAN SLEEP IN A TREE . . . OR NOT

In this chapter, I will cover some preferred characteristics for coops and runs. But chickens aren't picky and will likely be happy with whatever you're able to provide, as long as they have enough space and enrichment. And because I understand that the amount of money you have to spend doesn't equal the amount of love you have for your chickens, I will do my best to offer alternative ideas if conventional housing might be too expensive for you.

## How It Used to Be

I remember reading a comment in an online chicken group from a gentleman who lived in rural Alabama. He was an old-school chicken farmer of modest means who practiced what his parents practiced, who practiced what their parents practiced, and so on, for generations. The original discussion was about coops, and he volunteered rather innocently that his chickens slept in a tree at night and had no coop. Watching the other members of the group attack him in the subsequent

comments was like seeing hungry hens chase down a beetle.

"Don't you care if your chickens live or die?"

"You know there's such a thing as predators, right?"

"You really shouldn't keep chickens if you can't afford to care for them properly."

My experience growing up poor and yet surrounded by pets has taught me that someone's economic status should not affect their ability or right to enjoy the companionship of animals, certainly not when it comes to caring for chickens. Unfortunately, the above arguments seem to always boil down to what people can

Chickens feel perfectly at home in trees.

and can't afford. These habitually heated exchanges puzzle me because throughout history chickens have adapted well to the ingenuity of humans, even when those humans had little money to spend.

Although I do keep my chickens in a predator-proof coop and run, I think it's important to acknowledge that this isn't everyone's way and is not necessarily the "right" way. We live in a world with lots of variation—in climate, in culture, and in income level. As the gentleman in Alabama shows us, chickens are kept in a variety of dwellings—or no dwelling at all—across the United States and elsewhere. As long as chickens are kept in a manner that's appropriate for their climate and that keeps them generally safe, I believe every individual has as much right to keep chickens as I do, and certainly more right than any factory farm.

## Placing Your Coop and Run

I realize that many people don't have an array of options for where to place their coop and run. Optimally, the best place for your flock to live would be in an area of your yard that has great drainage. But even if you don't have a choice for where your coop will go, it is important to know that chickens create erosion over time. They spend their lives pecking and scratching at the earth and digging holes in which to dust bathe (chickens clean themselves by bathing in dry dirt rather than water). If there is vegetation within their reach, they will most likely demolish it in short order, depending on how much space they have. Their actions and movement in an area can literally change the landscape, so it's important to anticipate this regardless of where you place your chicken housing.

Depending on where you place your coop and run, the erosion your chickens create can lead to drainage problems that cause mud or flooding. I had some pretty intense difficulties with drainage in my first chicken run because I didn't think beforehand about how the chickens would affect the landscape. If I had been more thoughtful, I would have either placed my chickens in an area that was higher, with the run on a slight angle, or improved the drainage in that area as much as I could before I put my chickens there. I definitely would not have placed my coop and run at the lowest point of my yard! All this said, if a low point is your only available space, don't worry, all is not lost.

Here are some ways you can improve drainage and hopefully avoid a muddy mess.

- ✦ Regrade the area to add a slight slope where the run will be.

- ✦ Raise the run level with a substrate such as wood chips, straw, or a sand and gravel mix.

- ✦ Build trenches to lead water away from the coop and run area.

It is best to place your coop and run in an area with good drainage.

---

+ Add roofing to your run that has enough overhang to prevent excess water from getting in.

+ Add rain barrels to collect excess water for future use.

If there is standing water or mud in your chicken run, here's a nifty hack you can try until you fix your drainage issue. Sprinkle either pelleted horse bedding that is 100 percent pine or cat litter that is 100 percent pine onto problem areas to dry it out. Use it sparingly, as it will expand; you can always add more if needed. The pine pellets are highly absorbent and will soak up moisture in record time. You can rake it occasionally to prolong its absorbency power, and you shouldn't have to remove it since it eventually breaks down. It even helps with muddy chicken smells!

You can also use wooden pallets or even two-by-fours raised by bricks or concrete blocks to get your chickens out of muddy areas. The main thing you want to avoid is having your chickens constantly standing in mud or water. It's not good for them, and it can create smells your neighbors won't appreciate. Above all, make sure to plan ahead so you can prevent mud and floods. Hopefully, for your sake and your chickens', you can avoid these issues.

## BOTH SUN AND SHADE ARE IMPORTANT

Chickens love sunny spots in which to sun and dust bathe (more on this on page 126), which is very important for their health and happiness, but they also need to be able to get out of the direct sun, especially if they live in a hot climate. I have friends in Tucson, Arizona, who got their first flock of chickens in the early spring. Everything was going great until the 110-degree part of summer hit and they realized the sun was beating down on their small coop and run all day long. They were forced to hastily move their entire setup to the opposite end of the yard, where shade was provided by trees, and to add tarps. My heart went out to them, as I know the whole experience was terribly stressful. It could have been avoided had they considered how much sun their coop would get in the summer months, but in the world of chicken keeping, hindsight is often 20/20.

Though I live in Northwest Washington State, where summer temperatures usually stay in the 70s and 80s, it's still important for me to provide shade for my flock in addition to their coop. This is because, in general, even heat-hardy breeds don't do well in the constant glaring sun. I suggest observing the area where you plan to place your chicken yard to make sure there are appropriate amounts of both shade and sun throughout the year.

Of course, as we discussed before, many aspiring chicken keepers are working with limited space. If you find that the only place where you can put your chickens doesn't have any shade, well-placed tarps, sheets of plywood, or a small covered pen within their living space will work wonders. You can also plant chicken-friendly bushes and trees for some natural covered spaces. Just make sure that the roots are protected from any pecking and scratching!

## CLEAR THE AREA TO AVOID RODENTS AND SNAKES

As we discussed in the first chapter, many people associate chicken keeping with rats. In some ways, their concerns are not unfounded. There is no denying that a chicken yard can be an exceptionally attractive place for rodents, but there are certain actions you can take to make it less attractive so that they won't move in and start their furry family.

Rats and mice burrow and hide during the day and venture out at night. Tall grass, debris, woodpiles, or an old car or appliance are optimal nesting places for them to find comfort and safety. These are also good hiding places for snakes, who are notorious egg and chick eaters. Please note: I carry no judgment for "yard junk" whatsoever; when I was a kid, an old Plymouth Duster sat on our dirt driveway for several years (that was a fun car when it was running). All I'm saying is that you will definitely want to clean up any debris where you plan to place your chickens.

Otherwise, you're just asking for a rodent or snake issue down the road.

If you place your coop directly on the ground, rodents will find it most attractive to burrow beneath. If you have a smaller coop, it's best to raise it above the ground. Not only does this provide extra covered space for your chickens, but it also deters rats, mice, snakes, and even opossums and skunks. If your coop is large and must be on the ground, it's best to place it either on a concrete slab or on gravel. I recommend using at least 4 inches of ⅝-inch minus crushed gravel over landscape fabric. Make sure it's level and compacted.

---

Chickenlandia's coop is placed over 4 inches of compacted gravel.

## Important Considerations for an Easy-to-Clean Coop

Whether you are buying or building a coop, certain design aspects will make a huge difference in the amount of time you are able to enjoy with your chickens versus time spent struggling to keep up with their care. Trust me, while spending your days cleaning the coop or searching for eggs in cracks and crevices is fun at first, after a while, these chores become monotonous and even overwhelming. This is especially true if you experience a life change such as a new baby, a change in employment, or (heaven forbid) lessened mobility. I mention the last one not to be a bummer, but because it happened to me.

After I birthed our second child, I actually had to hire help so that I could keep up with the demands of my chicken coop and run. I tried to handle it alone, but I had chronic pain and limited mobility, so my coop ended up getting dirtier and more unkempt than I would like to admit. I even had one chicken die suddenly, and though looking back I'm sure it wasn't my fault, at the time I blamed myself because I was just so overwhelmed. Alas, life is unpredictable, and we can't always plan for the worst lest we live under a cloud of doom. But creating your coop and run with user-friendliness in mind can be a real lifesaver.

It is imperative that your chicken coop be easy to clean. Many prefabricated coops and coop plans are cute, but they require lots of bending and contorting to clean. I don't know about you, but my back cannot handle that kind of stress. I need to be able to either stand upright or sit on a chair while I'm shoveling shavings or scraping poop. I also don't want to be on my hands and knees in a dirty, poopy coop! The good news is, there are plenty of design strategies, even for smaller coops, that eliminate the need to bend over to do deep cleaning. Many of them use an under-roost pullout tray for extra ease.

Large walk-in coops can seem overwhelming when cleaning time comes around. In reality, if you use some of the same principles used in small coop design, the cleanup can be very manageable. Using trays under the roosts (the perches or branches that chickens sleep on through the night), for example, makes quick cleanup a breeze. And you can easily implement the deep litter method (see page 94), which requires less work overall.

## MAKE ALL COOP AREAS HUMAN-ACCESSIBLE

Back when I lived in the suburbs, I housed my chickens in a repurposed garden shed. It was a great little coop, but I didn't realize that it had an opening in the wall large enough for a small chicken to fall into. One day, after searching for one of my tiny chickens for what seemed like hours, I realized she was trapped inside the wall! I was able to rescue her, but I had to take apart the coop to do so. Needless to say,

you don't want this to happen to your chickens. Make sure there are no areas of your coop and run where a flock member can be out of your reach. Obviously, you wouldn't want your chickens to find themselves trapped and in danger of harming themselves or unable to reach food and water. But chickens will also hide when they don't feel well, and if you can't reach them, you won't be able to help them.

Another thing hens love to do is find hidden areas in which to lay eggs. If you have chickens that tend to go broody, you might find yourself with a broody hen sitting on 15 (or more!) eggs in a place where you can't access them. Even if you want baby chicks, it's important for them to be in a safe place and for you to be able to access them if needed. There are just so many reasons to make sure you examine your chicken coop and run well and imagine all possible scenarios. Doing so could save you a lot of time and possibly heartache in the future.

## GENEROUSLY ESTIMATE YOUR COOP SPACE

As we discussed in the previous chapter, each standard-size chicken needs 2 to 4 square feet of space in their coop. Only go with the lower number if your chickens have plenty of areas beyond their coop in which to find relief from the elements. One thing to be aware of is that most prefabricated coops make capacity claims that are too high, and even coop plans can sometimes overestimate the number of chickens they should hold.

If you abide by my guidelines for coop space and allow at least 10 square feet of run space per standard-size chicken, you will be giving your chickens the room they need to avoid problems associated with boredom, uncleanliness, or stress. When chickens are overcrowded, not only are they harder to clean up after, but they can also form bad habits such as feather picking or bullying, both of which can lead to a sick, stressed flock.

## SPACE YOUR ROOSTS, IF POSSIBLE

When choosing your type of roosts, you should take into consideration not only the comfort of your chickens but also your comfort when it comes time to clean them. Simple two-by-two boards, installed with the wide side up, are easy to clean, and chickens seem to like them. I like to paint mine with a gloss paint that is easy to wipe or hose down and scrub if needed. If you're able to make them removable, you can maintain them even more easily. If removability is not an option, scraping the roosts weekly with a paint scraper and wiping them down occasionally with equal parts water and white vinegar will do the trick.

Some people feel that roosts should always be flat because chickens prefer to sleep flat-footed. I prefer flat roosts because they are easier to clean, but the idea that round or natural branch roosts are "bad" for chickens perplexes me. Once again, I refer to how chickens have lived for millennia and how they would choose to live if they ever found themselves in the wild. They would absolutely sleep in the trees, and last time I looked, tree branches are not flat! If you do want to use round or natural roosts, make sure their diameter is large enough that chickens can cover their feet with their feathers, especially in colder climates. This can help to prevent frostbite on their feet and toes.

I recommend at least 10 inches of roost space per standard-size chicken. This takes into account nighttime flock dynamics, when more docile chickens near the bottom of the pecking order might need to roost away from the others. If possible, place the roosts at least 15 inches apart if they are at different heights, and 18 inches apart if they are at the same level. This will lessen the chances of chickens pooping on each other at night. Try to keep the roosts at least 10 inches from the coop wall, so the chickens won't soil the wall. Again, these are not hard-and-fast rules. I've seen lots of great coops, including high-quality ones, with roosts quite close together or close to the wall, and the chickens do just fine, although cleaning might be a tad more challenging.

Make sure you can access every roost in your coop. When chickens sleep at night, they go into a kind of comatose state. Flighty chickens that are utterly

Having enough roost space is important for avoiding pecking order issues at bedtime.

A good headlamp can be a lifesaver when dealing with chickens at night!

impossible to catch during the day are conveniently easy to pluck off a roost during nighttime hours. This comes in handy if one of your chickens becomes ill or injured, or if you just want to examine them as part of your chicken health routine. Imagine the ease of being able to go to your coop at night with a headlamp on and gently check over every chicken without them panicking into a cloud of feathers and dust!

## CHOOSE THE RIGHT SIZE AND NUMBER OF NESTING BOXES

Nesting boxes that are about 12 inches tall, 12 inches wide, and 12 inches deep can usually comfortably accommodate chickens of all shapes and sizes. You can go a little bigger if you have extra-large chickens like Jersey Giants, or smaller if you have itty-bitty chickens like Seramas or Old English Game Bantams. Place them in as private a spot as possible within

Nesting boxes don't need to be fancy! This one is a repurposed pet carrier.

your coop, or hang curtains on them to compensate for lack of privacy. I suggest supplying one nesting box per four or five chickens. Too few nesting boxes can create a stressful environment, with hormonal hens arguing over who gets the "best" box.

You can purchase or build your own nesting boxes, but rest assured that they do not need to be elaborate or expensive. It's perfectly suitable to fashion nesting boxes from items you already have lying around that otherwise would end up in a landfill. I've seen numerous imaginative nesting box creations, many of which are not only functional but also easy to clean. A few examples are 5-gallon buckets, milk crates, mail crates, or covered cat litter boxes. Since all these things are made of hard plastic, you clean them by simply taking them out of the coop and hosing them down.

It's normal for very young chickens to prefer sleeping in a comfy nesting box rather than on the roost, but they should outgrow this habit once they reach maturity. Some grown chickens, however, are determined to spend their nights on the nest. This can be a rather annoying habit for their caretakers since the chickens will always be soiling their boxes, resulting in dirty, poopy eggs (yuck!). If you have the space and your coop design allows for it, place the nesting boxes lower than the roosts. Hopefully, this will discourage the chickens from sleeping where they lay, since their instincts will draw them to the higher spot to sleep safely.

Some chickens—namely Silkies, Frizzles, and Showgirls—seem to want to sleep in their nesting boxes no matter what you do to deter their behavior. Chickens who are elderly, have special needs, or are chronically ill will also choose a low, warm spot to rest. Reusable nesting material, such as an old towel or blanket, can limit waste and ease cleanup. Other nesting box bedding options are straw, shavings, hemp, or disposable nesting pads. Some people will even use leaves, pine needles, or various soft foliage from their yard.

## The Importance of Ventilation

It's common for new chicken keepers to feel as though their chicken coop needs to be well insulated, with all nooks and crannies carefully sealed. After all, that's how we want our houses, right? But chicken coops are not houses. Since your flock lives in the same place where they make waste, they are under constant threat of ammonia fumes, which, if chronically

inhaled, can cause numerous issues. Good airflow is imperative, not only in the summer but in the winter as well. In fact, one of the main reasons why chickens end up with frostbitten combs, wattles, and toes is not low temperatures but rather moisture buildup from lack of ventilation.

The amount of ventilation you need will vary depending on the climate you are in, but a good rule of thumb is ½ inch of ventilation per square foot of coop space. Quality prefabricated coops usually come with ample ventilation, and any professional plans should also include it.

If you're converting a shed or other structure into a coop, it's easy to mindfully add ventilation (see the box below for some ideas). You want cross airflow that does not create a draft where your

---

### Ventilation Ideas

❖ Drill holes along roof peaks.

❖ Install windows with hardware mesh screens.

❖ Use wind turbine vents.

❖ Create roof cupolas.

❖ Install mechanical fan vents.

❖ Use wall-roof gaps.

❖ Add skylight vents.

Providing lots of predator-proof ventilation is essential for a healthy coop.

summer months hit. But always keep the ventilation that's at the top of the coop open year-round, even if it gets really cold. I know this is counterintuitive, but trust me, the cold is not your enemy as much as moisture is. Moisture buildup combined with ammonia fumes is a recipe for respiratory illness, among other unpleasant conditions.

To prevent critters from entering the coop, cover all ventilation openings with hardware mesh. Hardware mesh can be pricey if purchased new, but it's usually easy to find it used in online chicken communities or other classifieds. Please avoid securing vents with cheaper chicken wire if you can. If you absolutely must use chicken wire, double it up to make it less penetrable.

A quick internet search will yield various coop designs that are three-sided, meaning one wall of the coop is open-air and optimally covered with wire. Some chicken keepers say this method makes for more robust chickens with fewer health issues, especially respiratory, due to the high level of ventilation. My only hesitancy with this type of coop is that you must make sure you place it carefully and have a mindful design. You don't want heavy rain, snowdrifts, or drafts affecting your chickens. And if you have breeds that aren't perfect for your climate or you've decided to let your chickens reach old age, you might need to keep them warmer than you can in a three-sided coop.

chickens roost. An easy way to accomplish this is to place openings above the roosts so that drafts flow above the flock as they sleep. You can also add ventilation near the bottom of the coop that can be kept closed during winter and opened once the

## The Magic of Repurposing Materials

A wonderful way to bring costs down significantly when building your own flock housing is to use repurposed materials. A trip to Goodwill, a virtual peek in freecycle groups, or a search of online classifieds can yield some perfectly good materials in exchange for mere pennies or nothing at all. You can often find free wood pallets, an excellent source of lumber, by visiting your local hardware, grocery, or furniture store and asking if they have any to spare that would otherwise be discarded. Just be sure there is nothing about the material that could harm your chickens. Look for clean materials that are free of chemicals, mold, insects, rusty hardware, and other obvious damage.

At my previous property, a friend found a garden shed for me in my local classifieds, and I repurposed it into a gorgeous little coop. It was about 4 feet by 6 feet, with a tall ceiling and one big window. For roosts, I used an old ladder, a used wooden sawhorse, and some scrap wood. Due to lack of room, my nesting boxes had to be placed one above the other, two of which were higher than my roosts. When I started my YouTube channel, I would occasionally get comments about the poor design of my coop. "Your roosts are too crowded!" they would say. "Your nesting boxes are too high!" "I feel sorry for your chickens. I bet they hate living with you."

I confess that these comments would make me feel pretty insecure. I started to think that maybe I should stop teaching until I could show my audience a better-designed coop, one with the "right" roost and nest placement and newer materials, like all the other backyard chicken educators seemed to have.

Then I remembered Maria's chickens, sleeping in their designated tree under the Guatemalan moon. I realized my chickens were happy, healthy, and perfectly fine.

> ### Repurposed Materials for Coops
>
> ✤ repurposed shed or barn
>
> ✤ repurposed car port
>
> ✤ used wood from a torn-down fence
>
> ✤ wood pallets

# Chapinlandia

I want to take you back to one Saturday morning when I was around 14 years old. My dad, the ever-early riser, woke up before all of us and shuffled down the hallway in his well-worn slippers, just as he did every weekend morning of my childhood. He propped his shortwave radio at the end of our kitchen table, turned one of the enormous knobs until it clicked, and woke up the entire house with jarring static broadcasting from another land.

Finally, a signal formed.

"CHAPINLANDIA!" the announcer proclaimed as the haunting melody of a seven-member marimba band filled our home. The song was "Luna De Xelaju," which translates as "The Moon of Xelaju."

Xelaju is the K'iche' name for the city of Quetzaltenango, which lies in the mid-western highlands of Guatemala. The K'iche' Maya are Indigenous people who reside in that area, and there is some thought that the marimba originated with them.

As the music traveled across countries, a dreamy smile cast itself over my dad's face. *Chapin* is the colloquial term for Guatemalan people. So, *Chapinlandia* literally means "Chapin Land." I wandered sleepy-eyed into the kitchen and could see the relief of nostalgia in my dad's eyes. But his expressions were always accompanied by a lingering sadness. By the time I was born, he had been in the United States for almost a decade, but I

know he never truly felt like he belonged here. Years later, my mom revealed to me that he was bound in place by a vow; it was a promise he made to himself that he would not return to his homeland until the war was over and its dangers had passed. Sadly, the war dragged on for almost four decades. For all that time, my dad wanted to go home.

Like most teenagers, I went through a rocky relationship phase with my dad. But in the mornings when I would find him at the kitchen table, everything was always forgiven. Despite our respective shortcomings, my dad and I could not help but be dear friends and kindred spirits. We would talk for hours about the universe, politics, religion, and the meaning of life. That morning, after our usual philosophical musings, I once again told him that when I grew up, I was going to change the world.

"I'm going to be an actress, a writer, and a musician! And I'm going to help people and change the world!"

"Be careful," he replied solemnly, as he always did. "It's not that easy."

This is where a lot of our conversations ended at that time, with me feeling alienated and taking off to my best friend's house, and him calling after me, asking for us to spend more time together. He didn't understand why I was suddenly brooding and resentful, and I didn't have the words to tell him how much I needed him to believe in me. The barriers of culture and generations were just too much for us to break down. Of course, I never asked him about the dreams he was forced to let slip away. If I had, I would have known why he worried about me wishing for things that might be out of reach.

My best friend and I visited our special spot in South Side Park. We sat for hours in the clearing by the creek, making jokes about teenage things. She was a great friend—supportive, sarcastic, and a little bit tough. While we talked and laughed, she helped me to hold on to my dreams despite perennial seeds of doubt. There was no denying that we came from two drastically different worlds, but somehow, she and I were able to support each other. Maybe someday there could be a world where everyone did that for one another. If I could just hold on to my dreams, maybe someday I could help to make that happen.

The music of Chapinlandia played over and over in my head. In the distance, the rooster in that overgrown yard crowed.

# Living in Harmony with
# PREDATORS

Predators are a crucial part of any ecosystem. The best way to deal with them is to physically predator-proof your coop and run so that predator and prey can exist separately. I will cover supplies you might need to purchase, although I will offer lower-cost options when possible. Of course, there are many folks throughout history who never had a coop for their chickens and still managed to keep predators at bay by using dogs and other guardian animals. If you have a wonderful dog that is protective and gentle, you may not have to pay as much attention to predator proofing as long as your dog is on duty. It's also possible you have a wonderful dog that is not protective of chickens, but instead wants to have them for supper. I will cover that scenario as well!

## Predators Are Important

In the rain forest near the ancient Mayan ruins of Tikal, Guatemala, a curious sound echoes through the treetops as the first glimpse of twilight blankets the land. The chirping lingers somewhere between bird and monkey, yet it is alien enough from both to be mildly foreboding. Monkeys and birds don't stalk this way; they don't possess impossibly wide eyes that glow through the shadows of branches. No, this peculiar call comes from a 12-pound spotted cat with teddy bear ears and an elongated body designed perfectly for arboreal existence. To see this sprite of a feline is to be instantly enchanted. Known locally as a tigrillo or tree ocelot, the margay and its habitat are in danger of disappearing.

As she hops from tree to tree toward the forest's edge, the margay spots a group of huts with palm thatch roofs. One yard holds the scatterings of humans: an old motorcycle leaning against a stone wall, a pair of red shorts hung to dry outside a window, and, to the margay's delight, a small flock of domestic chickens pecking and scratching the healthy soil. Unlike savvy jungle critters, this prey is

Central and South America's delightful margay cat on the hunt

vulnerable and more oblivious to its surroundings, concentrating its attention on the ground rather than any dangers lurking above. This ill-fated evolution is certainly to the margay's advantage. Like an apparition, she rears up in preparation to swoop down and catch her prize.

But, alas, today is not her day.

A floppy dog comes trotting from the open door of his human's kitchen, a tortilla in his mouth. He positions himself protectively in the middle of the flock, then plops down to consume his treat from between large front paws. Knowing they are safe, the chickens begin their routine of finding a roost for the night. The margay, defeated, slips back into the shadowy jungle in search of her proper meal of rodents, insects, birds, or maybe a small monkey.

Luckily, margays are now protected throughout most of their habitat, which includes the jungles of Mexico, Central America, and South America, but deforestation remains their greatest threat. As an essential member of a delicate food web, if the margay were to go extinct, ripples from that loss would be felt across their extensive habitat and beyond. This is due to the special relationship between predator and prey and the interconnectedness all creatures on this planet share, including humans.

Without the margay, some prey populations would balloon to proportions optimal for the spread of disease. Rodents and

## A Note about the City and Suburbs

Many people living in the city or the suburbs assume that their chickens will be safe from predators because they don't see wildlife in their neighborhood. This is a critical mistake that could cost their chickens their lives. Even big cities like Chicago and Los Angeles have predators. In fact, two of the most dangerous ones are prevalent in densely populated areas: raccoons and domestic dogs. So please, no matter where you live, protect your chickens. There's something lurking that wants to eat them, even if you don't see it!

other pests would be more likely to infest villages and cities due to a lack of food supply. An imbalance due to the margay's extinction could literally change the way creatures large and small move throughout their territories, altering the landscape and even the very soil they walk on. At one point, humans seemed to understand that predators were a crucial part of an interconnected world, but we have lost some of this intuition.

Losing a chicken or chickens to a predator is a terrible feeling. I know this because it happened to me more than once. Yet, even though predators in my area are generally not endangered, I feel strongly that killing or even trapping them is not the best answer. Just like the margay, animals like bobcats, raccoons, weasels, and snakes play an important role in local food webs. Without their presence, rats and other rodents can over-multiply, creating an imbalance in the local ecosystem, which can lead to widespread disease.

Additionally, when rodents are allowed to overpopulate, they grow hungry from having to compete for food. This means it becomes more likely for them to arrive in your chicken yard looking for a meal and a safe spot to nest. Obviously, none of this is good for the rodents, your chickens, or you. Can you see now how predators might be important to the ecosystem in your backyard? Let's learn how to predator-proof your chicken yard so that predators and prey can continue to live separately.

## Types of Predators

New chicken keepers often feel overwhelmed when researching how to predator-proof their coop and run. Because so many different types of animals love the taste of chicken, the goal of deterring them can seem unattainable. I have found that grouping the predators into five different types can keep the overwhelm at a minimum. By concentrating on the behavior of each of these predator groups, you can make sure you're doing everything possible to keep your chickens safe.

### THE ONES THAT CREEP IN

There is nothing quite like being startled awake at 2:00 a.m. by the sound of a severely distressed flock. This is the rude awakening many new chicken keepers experience when they don't protect against sneaky nocturnal predators like raccoons, minks, foxes, and owls, to name a few. I call these predators that "creep in" because we often don't even realize they are among us until they attack.

From the very beginning, it is crucial that you instill the habit of locking up your chickens in their coop at night. It's best to close your coop right as dusk falls and open it in the morning *after* the twilight hours. Crepuscular animals, which include skunks, foxes, owls, and bobcats, are most active during this time.

For those of you who still enjoy a night on the town every once in a while (or every night—not my business!), an automatic coop door could be a godsend. There are several options, some more expensive than others. The cheapest option would be to buy the opener itself and attach it to your existing door. It's also possible to make one if you possess those engineering skills.

## PREDATORS THAT SQUEEZE IN

Perhaps the most challenging animals to keep out of your coop and run are those that have the uncanny ability to squeeze through impossibly small spaces. Rats may love pizza, but when it comes to cinching their waist enough to fit through a quarter-size hole, none can compare. Snakes can also get through small holes and crevices, and though they may not attack full-grown standard chickens unless you're somewhere with enormous snakes like Southeast Asia or the Everglades (yikes!), they will eat eggs, baby chicks, and perhaps even small bantams. Minks and weasels are horribly destructive and can easily slink through a chain-link fence and even holes 1 inch in diameter.

For so many reasons, the best material for chicken fencing and coop windows and vents is ½-inch hardware mesh. I prefer it over flimsy chicken wire (great for keeping chickens in but lousy at keeping predators out) or even chain-link fence. As I mentioned previously, look online for people selling or giving away free hardware mesh; you'll save a lot of money. If you don't have access to hardware mesh and are looking for alternatives, check out more money-saving ideas on page 62.

## THE TYPES THAT FLY IN

When I lived in the suburbs, I was under the impression that aerial predators were not a concern. For several years, my chickens and I enjoyed no contact with them.

Snakes love to snack on eggs and baby chicks.

But then one day I met a young Cooper's hawk who was determined to steal one of my bantams despite my embarrassing antics. I ran outside in my pajamas with a broom and dustpan, banging them together while screaming at the young hawk. (Sorry, neighbors!)

I was lucky that day, but I knew I had to take steps to protect my flock. Aerial predators, especially crows (who can be beneficial by warning adult flocks of predators but are known to steal both eggs and chicks), owls, and hawks, can be surprisingly bold when they have chicks to feed, are young themselves, or are just plain hungry. If you don't have a good farm dog or other large animal, the only surefire way to deter flying predators is a covered run. Roofing, strong bird-safe netting, or even chicken wire will suffice as long as your flock is locked up in a safe coop at night. Since raptors are known to have an aversion to shiny objects, you can try hanging tinsel, CDs, or predator tape around the chicken yard. I have found doing this to be very hit or miss, but it could buy you some time until you install roofing, netting, or wire.

## THOSE THAT CLIMB IN

Many creatures, including raccoons, opossums, and weasels, are avid climbers that can easily scale fencing if a tasty meal is on the other side. The same year the hawk visited, we had an unfortunate raccoon attack that took the lives of one very old chicken and two ducks, one with special needs. I felt terrible. This could have been avoided if I had kept my chickens in a covered run.

If you have a secure coop for your chickens at night, it's usually sufficient to cover your run with netting or chicken wire. Most animals that are strong enough to break through it will not do so in broad daylight, unless you live in a remote area and your chickens are fairly far away from your house. Of course, if you can't lock your chickens in a secure coop at night, it's best to use something stronger, like roofing or hardware mesh.

Electric chicken netting requires an initial investment but is an excellent deterrent of most land predators. You can move some brands from place to place in the yard to enjoy the bonus of providing your chickens with regular access to fresh grazing areas. You can also strategically place electric wires either near ground level or on the top of the fencing, depending on the kind of predators you are facing. Make sure the electric netting or wire you purchase is made specifically to protect chickens and be aware that open fencing will not stop aerial predators.

If you are living in an area with large predators, such as bears or mountain lions, your best option would be to either build heavy-duty electric fencing or have it installed. Of course, this is not cheap. But unfortunately, there is little else that would keep a determined bear or big cat at bay. It's possible that the other methods I've mentioned would buy you some time, however.

Electric chicken netting is the best defense against larger predators.

## THE ONES THAT DIG IN

Digging predators, such as coyotes and dogs, can come during the day and quickly dig a hole to access your flock. Domestic dogs can be especially vicious when their prey drive takes over, causing them to kill several chickens in a very short time. A good way to deter predators that dig is to either skirt (lay it down to create a border along the bottom of the fence) or bury hardware mesh or doubled-up chicken wire around the perimeter of your run.

If you skirt the mesh, place it about a foot out from your fence and make sure it's flush to the ground, using stakes if necessary. If you decide to bury hardware mesh or doubled-up chicken wire, place it about a foot deep. Mesh or wire that has been skirted or buried confuses animals, since they naturally dig at the bottom of the fence line. Going to such lengths may seem unnecessary, especially to those in the city, but I assure you that there are few predators worse than the domestic dog, and there are coyotes where you least expect them.

A coyote can dig under fencing to reach chickens in record time.

## Cost-Effective Predator-Proofing Ideas

Just as you can use repurposed materials to build your coop and run, you can find used predator-proofing materials. Every year, people jump on the chicken-keeping bandwagon, while others jump right off it (they must not have read this book)! A simple search through your local online classifieds will reveal used dog runs, fencing, wiring, and other building materials that can be used to safely secure your birds.

Here are some simple and relatively affordable ideas.

❖ used dog run
❖ repurposed hardware mesh
❖ doubled-up chicken wire
❖ used electric fencing
❖ repurposed fish netting (can sometimes be found for free at marinas)
❖ motion-sensor lights (serve as a deterrent and come in a range of prices)
❖ large rocks, bricks, cinder blocks, or buried wooden boards (use around the perimeter of your fencing)

---

### In Defense of Your Neighbor's Dog

I have four rescue dogs. As you can imagine, my love and compassion for dogs runs deep, and my heart aches every time I hear that someone's flock was attacked by their neighbor's dog or by a stray. It's a highly emotional situation, and I understand the tendency to be angry at the animal when they've hurt or killed your beloved pets. But before we think of ending the life of that dog (which I strongly disagree with), we need to step back and look at our predator-proofing.

The hard truth is that if a dog was able to get to your flock, it's because your flock was not protected enough. I know this is a tough pill to swallow, and please believe me when I tell you that I'm saying this with love. Yes, I understand that it's maddening when neighbors aren't responsible with their animals. But that doesn't lessen your responsibility.

Think about it: If a dog can get to your flock, a coyote, a raccoon, and a weasel can. Please don't blame the dog. Instead, understand that chicken keeping is full of lessons, and sometimes those lessons are hard. But the lessons are always for us, not innocent animals.

---

It's an unfortunate truth that sometimes going with cheaper options leaves your chickens more vulnerable than if you used more expensive items. For example, chicken wire, even doubled up, will never be as strong as hardware mesh, and, as I mentioned before, a chain-link dog run does not deter predators that can squeeze through it. If your situation leaves your chickens in a more vulnerable state than you would like, choose to focus on creating as many barriers as possible, so that it takes longer for predators to get in.

## THE AMMONIA HACK

I learned this trick from our local wildlife center after I called them seeking guidance regarding pesky raccoons. Most

Get creative when setting up barriers between predators and your flock. Rocks, bricks, cinder blocks, and inexpensive motion sensor lights could save your chickens—and your pocketbook!

predators are territorial and will mark an area with their scent to keep the competition away. You can mimic this strategy using items you likely already have around the house. Here's what you will need.

- ✤ face mask
- ✤ rubber gloves
- ✤ bottle of ammonia
- ✤ bucket
- ✤ a few old rags
- ✤ some strong string, fencing wire, or rubber bands

Simply put on your face mask and gloves, then fill the bucket with enough ammonia to soak the old rags. Submerge the rags in the ammonia, then wring them out so that they are not dripping. Hang the rags just outside the perimeter of your chicken yard using string, wiring, or rubber bands. Don't put any rags near the coop or anyplace where the ammonia fumes could reach the flock, because inhaling ammonia is bad for chickens.

You will need to dampen the rags about once a week or if they get soaked by the rain, but be advised that this hack is not meant to be a long-term solution. When my chickens were attacked, I knew the critter that did the deed would be back for another easy meal, and since I wasn't able to put up netting right away, the ammonia trick really helped. It is not a foolproof tactic; animals are smart and can usually figure out our tricks eventually. But if you need to buy yourself some time, it can certainly do so. And if you want a little extra protection, you could also play talk radio in the chicken yard as a deterrent—another tip I learned from the wildlife center. Considering how obnoxious most talk radio can be, I don't blame the animals for not coming around!

## DON'T GIVE THEM A REASON

Did you know that some predators initially come around because they smell something other than your chickens? Imagine a racoon's delight at discovering a compost

Store all feed in secure, rodent-proof containers to discourage predators looking for an easy meal.

pile and then realizing that it comes with a chicken dinner.

If you are keeping your chicken feed container in the coop or run, and certainly if you are tossing your flock kitchen scraps or keeping a compost pile in the run, you *must* predator-proof these areas. Keep chicken feed in a critter-proof container or in an area that cannot be easily breached. If you can't purchase or don't have access to an appropriate feed container, you can make one by using bungee cords (see photo above).

I think it's a great idea to mindfully give your chickens kitchen scraps, but it's super important not to leave scraps lying

Serving up kitchen scraps and compost in a predator-proof container keeps your flock safer.

around the chicken yard for too long, especially overnight. There are two ways you can prevent this from happening. You could give your flock only what you know they will consume by dusk. This takes a little trial and error, and you may find yourself cleaning up half-eaten scraps before bedtime, at least initially. Or you could create a predator- and rodent-proof feeding station by building or purchasing an enclosed box with a good lid.

Even if you keep your kitchen scraps in a predator-proofed feeding station, you should be aware that most chicken-hungry critters have an uncanny ability to smell delicious things. This means their noses will bring them that much closer to your chickens. If you plan on feeding your chickens kitchen scraps or keeping a compost bin near your coop or run, you must take the appropriate precautions to make sure your chickens are adequately protected.

One final word before we move on: If the last two chapters have you feeling overwhelmed due to the cost factor (even cost-effective options are out of reach for some), I still think you should get chickens. It's possible they might not be as secure as you would like, but chickens face a certain level of risk just being on the planet. The benefit they will give you and your family must always be a priority in your equation. So do your best, and don't focus on your limitations. The quality of life you give your chickens will be miles better than that of most chickens, and that means something.

## The Joys of Chicken

# MOTHERHOOD

A chick's short life before she is introduced to you is usually pretty stressful. Chicks typically hatch out in a huge hatchery incubator, undergo a rather uncomfortable examination, and then are packed up and sent on a wild trip through the postal service to a farm store, where they are handled by giant hands and ogled before going home with you. Hatching in and of itself is an arduous process that they need time to decompress from—a trip across the country hardly helps.

In Chapter 8, we will discuss some ways that you can stave off any ill effects such a stressful beginning might bring forth. For now, know that once you acquire your baby chicks, doing things to support their natural instincts will give them the best chance at a balanced, issue-free life. For this reason, I encourage you to act as a mother hen would as much as possible. So let's think about all the things a mama hen would do to keep her children safe, healthy, and resilient.

## Mimicking Mama's Instincts

When a mother hen hatches out her baby chicks, she waits patiently until all viable eggs hatch, which usually takes about two days. During that time, the already hatched chicks dwell beneath her, peeking out through the comfort and safety of their mother's warm feathers while they wait for the arrival of their siblings. They can go without food and water during this time because for about 48 hours after hatch, they continue to absorb the remaining nutrition supplied to them from within their egg. This is why day-old baby chicks that are shipped through the mail are able to survive the trip. It's also why it's super important to get them to food and water as soon as they arrive at their destination.

Once all her chicks are hatched and fluffy, Mama Hen takes them out into the world. She calls them over with a clicking sound when she finds a tasty morsel, which could be chicken feed, an insect, a blade of grass, or even a tiny pebble to aid their digestion. As they move throughout the chicken yard, they are exposed to everything in their environment, including bacteria, viruses, fungi, and parasites. Even so, they are likely to thrive so long as their mother's humans have good husbandry practices in place.

When we observe a chick's natural life, we see that Mother Hen is supplying shelter, warmth, and safety; the ability to find

food, water, and grit (sand and small pebbles); and safe exposure to the wonderful world of nature. Let's go through how you can mimic this process.

## SHELTER

As we discussed in Chapter 2, when a mama hen cares for her chicks, it's called brooding. When a human is caring for baby chicks, they keep them in what's called a brooder or brooder box for as long as supplemental heat is needed. In general, you want at least 1 square foot of brooder space per standard-size chick, with the understanding that soon they will also need time outside as well as other forms of enrichment within their space such as a pie dish with dirt to dust bathe in, a clump of dirt and grass, and/or a small perch to practice roosting.

There are few problems associated with having too much space for chicks as long as they are safe from predators, drafts, and have a good heat source. But there are numerous issues that can arise when chicks are crowded. My suggestion is to overestimate how much room they need to reduce the chances of having any stress-related issues.

A very popular brooder is a plastic bin, like the type you would find in the storage section at your local department store. Most people have them, and for a few day-old baby chicks, it's perfect! But be aware that the chicks will grow out of this container surprisingly quickly, unless you have two or three exceptionally small bantams. They will also eventually gain the ability to fly out of it. For these reasons, I would either start out with something larger or make sure you have another brooder they can grow into.

My two favorite types of brooders are large guinea pig cages and galvanized metal tubs designed for feeding and

---

### Chicken-Care Supplies

**Shelter (brooder):** large plastic tub, large wooden box, guinea pig cage, galvanized metal tank

**Warmth:** heat lamp (red), ceramic heat bulb, radiant brooder (preferred)

**Water:** plastic or glass chick waterer, nipple waterer, small dish

**Safety:** additional items for waterers such as marbles or clean, smooth rocks

**Feed:** chick starter feed (non-medicated preferred), treats

**Feeder:** plastic, metal, or glass chick feeder or trough; small dish

**Digestive supplements:** chick grit or coarse sand

**Bedding:** paper towels, pine or aspen shavings, pine pellets, old towels

For the first few weeks, chicks need to live in a safe, temperature-controlled environment.

watering cows or horses. You should be able to use these from day one to week eight, depending on how many chicks you have. Both of these brooders can sometimes be found either used or for free in online classifieds. Just make sure you disinfect them with a little white vinegar and water before use. A covered guinea pig cage is handy because curious baby chicks can't fly out of it. Galvanized tubs are a great size for housing chicks, but you will need to lay some wire over the top to keep fliers inside.

Of course, if you are the industrious type, you could build your own chick brooder. In this case, be sure to construct it so as to limit fire hazards, allow for ample ventilation, and keep chicks in the brooder where they belong. A simple square or rectangular wooden box with a wire lid should do the trick. Just remember that if you're going to use a heat lamp, it must be secured properly and at least 6 inches away from the brooder walls while also allowing enough room for chicks to be completely out from under it and its heat if they wish to be. Make sure your brooder is easy to clean and has enough space for the number of chicks you are planning for. (And then add some

to that number, because there's a good chance you will get more chicks than you thought you would. Trust me on this!)

## WARMTH

When they're not able to snuggle under their mother, baby chicks need a source of heat to keep warm and cozy. Back in the day and even currently among some old-school farmers, chicks would be placed in a small box fashioned to hold in their own body heat. If more heat was needed, people would place hot-water bottles or even containers filled with hot water and then wrapped in towels or blankets in the brooder to keep chicks from getting chilled. These are both valid ways

Your brooder needs a lid. Little velociraptors (a.k.a. baby chicks) will eventually fly from open brooders. The last thing you need is them pooping all over your house! Plus, they may not be able to get back in and could get chilled.

---

of caring for baby chicks, although I do think the chicks could be more vulnerable to stress-related issues. Today in the United States, the most common way of artificially brooding chicks is to use either a heat lamp or a radiant heat plate. While the heat lamp is more popular and costs less, it is not my first choice due to the fire hazard it presents.

TOO COLD

TOO WARM

JUST RIGHT

Baby chicks generate a lot of dust, and heat lamps get really, *really* hot. A thick layer of dust on a hot bulb can ignite and cause a fire. Chicks also require the use of some kind of bedding, usually shavings. Imagine what could happen if a poorly secured bulb fell into dusty, dry shavings. This scenario is never good. Heat lamps have even been known to spontaneously explode! Honestly, I hate speaking negatively about heat lamps, because they are very affordable. But since they are cheap, they are also cheaply made. I cannot in good conscience recommend them.

My favorite heat source for baby chicks is the radiant heat plate, which is basically an electric panel heater placed horizontally on four adjustable legs that biddies (baby chicks) can get warm under as they would beneath a mother hen. The heat panel does not get hot to the touch, but rather warms the chicks' backs in the same way a mother hen's body would.

I love this device so much because I have witnessed firsthand how chicks raised under one are healthier than those raised with a heat lamp. The first advantage is that the chicks are not exposed to artificial light 24/7. Not having the natural cycle of day and night can be stressful for a chick, especially if a white bulb is used rather than a red one, because the red at least casts a warmer glow and so is not as much of a disruption to their natural rhythms. I like to place my chicks next to a window so that they can enjoy a natural dusk and develop their circadian rhythm, which in turn makes it easier for them to transition into living outside.

Baby chicks raised with a radiant brooder get to practice what is likely an important instinct: running toward and away from heat as needed. I don't have any hard-and-fast evidence as to why this is probably good for baby chicks, but I do think it's possible there are benefits to practicing these instincts that we don't completely understand. In my experience, baby chicks raised with a radiant brooder appear to experience fewer stress-associated problems such as starve-out, pasty butt (both are health problems that I will cover in Chapter 9), or cannibalism. It makes sense that this more natural beginning gives them a better chance of leading a life with fewer issues in general.

All that said, radiant brooders are expensive and therefore out of reach for many people. If you can't afford a new one, can't find a secondhand one, or aren't able to find one at all in your area, please use a heat lamp without guilt. Always choose a red bulb, secure it very well, and keep it free of dust. Observe your chicks' behavior—not the ambient temperature—to see whether they are comfortable. Are your chicks huddled under the lamp, peeping loudly, trying to sleep beneath or atop each other? They are too cold. Are they all lined up against the walls of the brooder, lethargic, or panting? They are too hot. Your baby chicks should be peeping softly, pecking and scratching, sleeping, and just . . . vibing (for you Gen Xers such as myself out there, that means they are just chillin').

## SAFETY

When you read the list of items you need for baby chicks, you may see "marbles" and think that I've lost mine! I promise you my marbles are still in place. Marbles or clean, smooth rocks made the list because when placed in the base of a fountain waterer or small bowl, they prevent baby chicks from immersing themselves. The chicks will sip the water that comes up through the marbles or rocks but won't be able to get in

To prevent drowning, place clean marbles or rocks in your baby chicks' water dish.

there and get themselves into big trouble. A drowning chick is obviously a terrible situation, but a chilled chick is also in grave danger. You may wonder why a baby chick would immerse themselves in water. Perhaps without the constant presence of a mother hen they just don't know any better, although it can also happen with a mother hen around, just not as often. Fortunately, it's easy to prevent.

The marbles can be removed from their waterers after a few days, when the chicks are bigger, more stable, and have gained enough awareness to stay out of the water. It may take a little longer for very small bantams.

## WATER

The most common chick waterers for small flocks are quart- or gallon-size fountains with a screw-on base. Fountain waterers work through the force of gravity; you fill the jug component with water, which then empties water into the base as baby chicks or chickens drink. The jug can be glass or plastic, and the base plastic or galvanized metal. Some are compatible with a quart-size glass jar.

You have several options for providing clean water to your flock, from a simple gravity-fed version (top) to a nipple waterer (middle) or cup waterer (bottom).

There are also nipple waterers, which work by releasing drops of liquid when pecked at, and cup waterers, which contain either a nipple or other trigger mechanism inside a small cup that the chickens can peck at to release water. Some nipple and cup waterers come with a jug, and others are just the cup or the nipple, which must then be attached to your own tub or bucket. For my own flock, mainly because I also keep ducks, I prefer the fountain type, in which the chicks can fill their beaks and then tip their heads back to drink. Chicks will, however, generally take to nipple waterers very quickly, and this type of waterer is easier to keep clean than the fountain variety. If you go the nipple or cup waterer route, just be mindful to watch that all your chicks are drinking as they should. If any of them aren't catching on, I suggest switching to a fountain waterer.

It is possible to use just a small bowl for your chicks' water, and in fact, that is what people used before chick waterers were manufactured on a large scale and it is still what many people use in places where specialized waterers are not available or not affordable. Make sure the chicks can't tip the waterer over or immerse themselves in it and drown or get chilled.

---

### Your Optimal Baby Chick Setup

- ✤ secure brooder box or cage, set up and ready in an area that is draft-free, does not get below 50°F (10°C), and is safe from predators

- ✤ two layers of paper towels or a thin, cloth towel flat on the brooder floor as bedding

- ✤ heat source assembled, safely secured, and turned on

- ✤ waterer with marbles or clean, smooth rocks for protection, filled with tepid water (see Chapter 8 for optional immune-boosting additives); water must not be placed under the heat source

- ✤ feed and chick grit scattered about the brooder so chicks can easily locate it

- ✤ feed with a sprinkle of grit in the feeder, placed away from the heat source

## FOOD

Baby chicks need chick starter feed (ration specially formulated for their stage of life) for at least the first 8 weeks of life. After that, some people transition them to grower feed and then to layer feed when they reach adulthood. I know that purchasing three different feeds for this short period of time is not an option for many people, and that's fine. It is perfectly acceptable to keep them on starter feed until they reach the point of laying; it won't harm your chickens in any significant way.

Starter feed comes in crumble or mash. The crumble is processed feed that contains the nutrients baby chicks need to grow and thrive. Mash feed is crushed or ground grain plus other vitamins and minerals. It is usually not processed. Chick starter can be conventional, organic, non-GMO, medicated, or non-medicated. Organic or non-GMO is preferable, but

---

A gravity feeder provides an ongoing supply of food while keeping chicks from pooping in and otherwise wasting it.

I recommend using the highest quality non-medicated feed you can afford. Medicated feed contains amprolium, which prevents a common intestinal disease called coccidiosis. I don't recommend it for backyard flocks, and I'll discuss why in Chapter 9.

When baby chicks come home to live with you, it's important that they have immediate, easy access to their feed. The best way to ensure they can readily find it is to sprinkle it around the floor of the brooder in addition to putting it in a feeder. Baby chicks, especially those that were born at a hatchery and shipped through the mail, are extremely vulnerable to a condition called starve-out. Starve-out means that they did not get food and water soon enough after hatching and are thus failing to thrive. They will stand alone, look listless and droopy, and not eat or drink. Starve-out is fatal if not caught in time and is best avoided by taking every precaution so that your chicks are eating and drinking as soon as possible.

While some chicken educators and enthusiasts feel that chicks should only receive starter feed, I like to supplement with other foods. Given that baby chicks would have a buffet of bugs, foliage, and whatever else Mama Hen encourages them to try in their natural environment, I feel that chicks should be offered a variety of safe food from the beginning. When my chicks come home, I sprinkle some crumbled hardboiled egg yolk with their feed on the floor of the brooder. Offering them yolk might seem to some like cannibalism, but it simply gives them an extra boost of nutrition that they already had when they were growing in their egg. As they grow, I will also offer baby chicks dried herbs such as oregano and thyme, crumbled grubs or mealworms, and vegetable scraps to nibble on and play with. We will discuss more about nutrition in Chapter 7.

## Feeders

The most common type of feeder for baby chicks and chickens is the gravity feeder, which has a food storage container on top that remains open to a food dish on the bottom. It works through the power of gravity, automatically refilling the dish as food is eaten. Gravity feeders come in various sizes and can either be completely plastic, have a plastic base with a glass container top, or be made entirely of galvanized metal. There are also trough-type feeders. Some gravity and trough feeders have holes that allow the chicks to eat the feed but prevent them from walking in and ultimately pooping on their food.

Just as you can use a small bowl for water, you can also use a shallow dish for feed. The only problem is that they just don't know not to poop in it! So if you feed them from an open container this way, check it often to be sure it's clean and free of droppings.

### Digestive Supplements

It is important to focus on your chicks' digestive health, and chick grit is a crucial part of keeping them healthy. Chick grit is basically coarse sand that, once consumed, goes into the chicks' gizzard and serves as "teeth" to chew their food. Chicks are born with the instinct to eat it when it is made available to them.

Grit is very important when baby chicks are little. Much of their survival depends on whether food is properly moving through them, so that they can absorb the nutrients their bodies need. Some chicken enthusiasts and educators might tell you that if your chicks only consume chick starter, you should not add grit to their diet until they get older and their diet becomes more varied. I believe in offering chicks a variety of foods from the beginning, so I think that access to grit is important from day one, especially if your flock is consuming raw starter feed.

Even if you put them on a strict diet of starter feed, your little *T. rex* descendants cannot be trusted to stay away from other, heartier fare. Lord help the random fly or spider that finds itself in your young flock's presence. It will be eaten in short order! The grit will help the chicks digest it properly. It's easy to access and easy to give to them, so why not? Sprinkle the grit about the brooder and into their feed container. You can buy commercial chick grit or give them clean, coarse sand.

Chick grit is coarse sand that goes into the gizzard to aid digestion.

### COMFORT

Baby chicks need some type of substrate on the floor of their brooder, both to help keep them cozy and to absorb moisture and odor from their droppings. When I bring day-old chicks home, my favorite beginner beddings are paper towels or a very thin, old towel (make sure it hasn't been washed in anything fragrant and has no strings hanging from it). This type of bedding offers excellent traction, which is very important for the proper development of tiny legs, feet, and toes. Young chicks can sometimes hatch with a predisposition for leg and foot problems due to genetics, vitamin deficiency, or an issue with incubation. Other times, chicks develop problems due to poor traction in their living quarters. It's best to lower the likelihood of these unwanted conditions by offering them good support from the beginning.

Because it is so important for baby chicks to begin eating as soon as they arrive in their new home, you want to be sure it's easy for them to find their food. Feed sprinkled on thin paper or cloth towels is especially easy for baby chicks to find and pick up. When new chicks can easily find their feed, they have the best chance to thrive and stay motivated to peck and scratch for their meals.

If you're using reusable cloth towels, make sure to change them out when they are soiled. You can keep chicks on these towels the entire time they are in the brooder, as long as you can keep up with cleaning them. I don't recommend keeping chicks on paper towels through their whole brooder experience, though,

Shavings are a good choice of bedding for slightly older chicks.

because as the chicks grow they will tear the paper towels to shreds, the brooder will become difficult to clean, and the towels will create extra waste. If your chicks are on paper towels or you grow tired of cleaning cloth towels, you can switch to pine or aspen shavings, pine pellets, or hemp bedding once the chicks have made it past the first few days at home, are

eating out of their feed dish, and appear to be thriving. Although some of my older farmer friends swear by cedar shavings, there is reason to believe they could cause damage to a chicken's sensitive respiratory system, so you may want to avoid them.

Clean out the shavings when the brooder is soiled and if there is any hint of ammonia, which can cause a whole slew

## Optional Items

Domestic baby chicks and chickens have survived for thousands of years with very little intervention from humans, so please try not to get caught up in feeling you must purchase numerous bells and whistles to take proper care of them. That said, there are a few items that I feel are pretty important even though they are technically optional, as they are beyond the basics of food, water, warmth, and shelter.

These extras are fairly affordable, and some you can grow yourself or you likely already have lying around the house. Since these items have to do with wellness and aren't necessities, I describe them in detail in Chapter 8. You may want to skip ahead if you wish

to compile your baby chick checklist right now, with these optional items included.

✦ electrolyte, vitamin, probiotic (EVP) supplement for chicks

✦ apple cider vinegar (ACV)

✦ dried herbs (oregano and thyme)

✦ green tea bags

✦ concentrated vitamin supplement for chicks, such as Nutri-Drench

✦ needleless syringe

✦ calming homeopathic flower essence (e.g., Bach Rescue Remedy)

✦ blow dryer with a low setting

of problems. How often you need to clean your brooder will depend on how much space your chicks have and how many chicks are in that space. Using common sense and your sense of smell will soon make you an expert on brooder maintenance! The main thing you want to do is keep the brooder dry and ammonia-free.

## EXPOSURE TO NATURE

In Chapter 8, we'll discuss the ways in which outside exposure aids your baby chicks' health and happiness. For now, know that if you're hand-raising your baby chicks, it's a good idea to expose them to the outside world when they are young and still in the brooder, just like a mother hen would.

At 2 weeks old (you can start sooner if the weather is warm), your baby chicks can go outside under your supervision on sunny days. Bring them to a location where there is both dirt and grass, and have fun watching them peck and scratch and even dust bathe just like adult chickens! Make sure they are in a secure area, and stay outside with them so that you can observe their behavior. Bring them inside if they start to get cold (they will try to get under each other, appear distressed,

or start to peep loudly), and please use common sense (sunny and 50°F [10°C] is very different from sunny and 15°F [−9°C]). Remember, you are their mother!

If the weather outside doesn't permit a field trip for your baby flock, bring nature to them. Grab a nice clump of dirt and grass (make sure it is free of pesticides or fertilizers and hasn't been heavily soiled by your existing flock) and place it in their brooder. They might be suspicious or afraid of it at first, but soon their curiosity

## A Chick's Road to Living Outside Full-Time

**Weeks 1 and 2:** Most of the time is spent in the brooder, with occasional trips outside if weather permits.

**Week 3:** Allow more time outside, and pay special attention to brooder enrichment.

**Week 4:** Chicks can work up to being outside for hours in temps above 70°F (21°C), but watch them closely to prevent them from getting chilled. Brooder enrichment is a must.

**Week 6:** Standard-size birds are usually ready to move outside permanently if the weather

permits. Bantams are usually able to be outside most of the day if weather permits.

**Week 8:** Almost all breeds are fully feathered and able to be outside permanently in weather above 50°F (10°C). If temperatures get below 50°F, acclimate slowly up to week 12.

**Week 12:** Most chicks are acclimated to all weather. Supplemental heat may still be necessary for extreme climates.

will take over and they will play with it, peck and scratch at it, eat it, and even dust bathe in it. This is a wonderful way to keep boredom down, and it will also give them an opportunity to build resilience to the outside environment.

You can also grab some dry dirt from your yard and offer it to your chicks in a pie dish for a mini-dust bath (more on dust baths on page 127). It will help to keep them clean and parasite-free as well as allow them to practice their instinct to dust bathe. Bonus: It's super fun to watch them bathe! You will know what I mean once you witness it for the first time.

## For Mama Hen, Timing Is Everything

Even though enthusiasts, hatcheries, and breeders can hatch eggs in an incubator (an enclosed environment in which to artificially hatch out eggs) all year round, there's a good reason why mother hens normally go broody in the spring. Baby chicks are usually ready to be outside permanently when they are fully feathered, which most often occurs between 6 and 8 weeks. If you are in a hot climate, they may be able to live outside much sooner. The opposite is true if you are in a very cold climate.

The handy thing about a mother hen is that she knows how to gradually acclimate her brood to the elements, and she won't go broody during a time when it would be too cold to hatch babies. (There can be exceptions to this; some chickens apparently can't tell time!) To reduce stress for both you and your new little flock, get your chicks during baby chick season, which is generally considered February through May or June, with careful consideration of your local climate.

When you raise chicks in a brooder, it's extremely important to consider what your local climate will be when the chicks are fully feathered at around 8 weeks. If day or nighttime temperatures are below 50°F (10°C) when you first bring out your feathered chicks, you may need to slowly acclimate your flock to the outside by either bringing them inside when needed or offering safe supplemental heat, like a panel heater, outside during the transition.

When you put your baby chicks outside, it's important for them to understand exactly where they are going to live. You can't just throw them in the backyard near the coop and expect them to recognize their home! To prevent them from getting confused and wandering off, it's best to either keep them inside the coop for a few days (if coop size permits) or let them spend the day outside in an enclosed run that is attached to the coop. This way, they can't wander too far, and hopefully they will go into the coop at dusk. Since your baby chicks were not raised by a mother hen, their instinct to roost or even go into the coop may be delayed. It's not a bad idea to check on them to be sure they've gone to bed. You can also place them on their roost at this point if necessary. Don't worry, they will get the hang of it!

## You're Ready

Once you bring your baby chicks home and get them into their new digs, you will understand just how important it is to have everything set up beforehand and you will be so thankful you did. But the preparation is usually the toughest part. Your chicks' day-to-day care from that point on will likely feel like second nature to you. After all, humans have been doing this for a long time. The most important things for you to do for the next 8 weeks will be to remember the basics of mindful chicken rearing:

* Pay close attention to your little flock's behavior.
* Make sure they have access to warmth.
* Keep their brooder reasonably clean.
* Ensure they always have food, water, and grit.
* Above all else, use your common sense.

If you come up against something that is worrisome, it's okay to ask your chosen chicken community for advice. But remember that if certain suggestions don't feel right to you, they probably aren't. I liken chicken rearing to having children: No one really tells you everything you're supposed to do, but somehow you already know. Even if you feel like you don't.

# DAY-TO-DAY
## Care for Adult Chickens

Raising baby chicks is such a gratifying experience. At the point that your biddies have become teenagers, the feeling of accomplishment and confidence in your chicken-keeping abilities should be deservedly high. Think about it: You raised the vulnerable young of another species from newborn to adolescent and not only permitted their survival but also gave them the opportunity to thrive. So please do me a favor and lift your right hand, reach up over your shoulder, and give yourself a nice pat on the back. You are an established "good chicken parent." Congratulations! Now it's time to enjoy your chickens, who need less day-to-day care than you might think.

## A Chicken Keeper's Morning

It's 6:00 a.m. and the sun is peeking over the horizon. If you have a rooster, he may have already been belting out the occasional crow for an hour or two. His melody rings in a new morning, a new opportunity to connect with nature, and a chance to leave the concerns of yesterday behind. If your coop is large enough, you could hit the snooze button more than once. But if your coop only allows for roosting space, it's time to get up and let your flock out to stretch their legs and wings.

You roll out of bed. There's no need to change your clothing; you already have on the official chicken keeper's uniform (pajamas and robe). You tie an apron with pockets around your waist, and if you're fermenting feed (we'll talk about this in the next chapter), you grab today's batch off the counter. You slip on your muck boots and head out the back door.

You can hear the clucking coming from inside the coop, but you don't open it yet, preferring to complete your chores without curious-bird interference. Your first action is to rinse out and refresh the water containers. If you're using nipple waterers, you can skip the rinsing on most days and just make sure they are filled with fresh, clean water. Next, you rinse out and fill their feed dish with new ferment, or simply make sure their feed container is topped off if you are free-feeding. You also make certain their calcium supplement dish is full and scatter some layer grit about the run.

It's time to let the flock out. As soon as their little door opens, the chickens rush to their feed and water bowls. It's early, but just in case, you check each nesting box. To your surprise, you realize the ladies have already been busy. You find four eggs, still warm, and pop them into your apron pockets.

After spending a couple of minutes tidying up the roosts with your paint scraper, you are done with your morning chores. You could go back inside right now and begin your chaotic day, but instead you sit in your special chair for a few fleeting moments, listening to the soft coos and clucks, soaking in the peace of your flock. Many before you have done the same, and hopefully, many will after.

## Feeding and Watering

I'll cover chicken nutrition and feeding options at length in the next chapter, but for now, let's chat about commercial chicken feed, which is the most common diet for chickens in the United States and other Western countries. By the time your hens reach maturity, they will need the nutrients contained in a feed formulated to meet their bodies' reproductive demands. You can start them on layer feed when any flock member starts laying (usually 16 to 20 weeks, depending on breed and time of year), or, if no chickens have started laying by 20 weeks, start them on layer feed at that time.

You can purchase layer feed with varying levels of protein, but 16 percent protein should meet a beginner flock's needs just fine. The three main types of layer feed are crumble, pellet, and mash.

**Crumble and pellet** are processed feeds, and though there's really no difference between them other than size, you may find that your chickens prefer one and stick their beaks up at the other. Luckily, they both offer the same benefits.

**Mash feeds** are most often raw and unprocessed, and contain whole or crushed grains with other important nutrients mixed in, usually in powder form.

If your flock is on mash feed, you may want to soak or ferment it to ensure that your chickens are eating all the components, rather than picking out the parts they prefer.

Most folks who choose pellet or crumble types free-feed their chickens. This means they give their flock access to food all day, usually in a gravity-type feeder. In general, a standard-size chicken will consume about one-half cup of dry feed daily, but this will vary depending on the season they are in. I find that my chickens eat the most during spring, when they are laying in full force, whereas they don't seem to eat as much in the winter, when the days are shorter and they aren't laying.

CRUMBLE

PELLET

MASH

I don't free-feed my chickens, but rather give them their ration of fermented feed in the morning. By fermenting their feed, I am able to save money while increasing the nutritional value of the feed, which I'll explain in the next chapter.

**A calcium supplement** should be on offer to laying hens at all times. You can use oyster shell, the hens' own crushed and clean eggshells, or limestone calcium supplements. Simply free-feed it to them in a separate dish and make sure to refill it as needed.

**Grit** helps your chickens digest their feed, and they will need it for the entirety of their lives. As discussed earlier, baby chicks and chickens instinctively know that they must find and consume coarse sand, small rocks, and pebbles, which then travel to their gizzard to break down their food. Chickens with access to grit can properly digest a variety of foods, including all manner of plants, seeds, grains, and insects. Chickens on pasture can usually find the crushed rock and pebbles they need within their natural environment, but if your chickens live in an enclosed run, you will need to offer them consistent access to a commercial layer grit, which is basically crushed rock. Scatter a handful or two of grit about the run as needed, or offer it at all times in a separate dish.

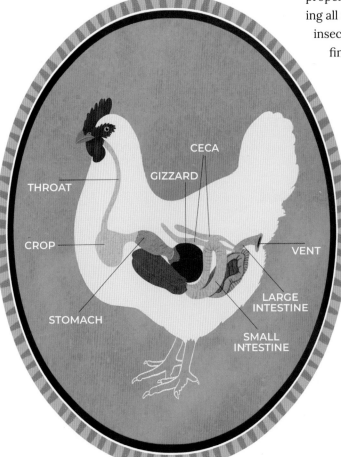

THROAT

CROP

STOMACH

CECA

GIZZARD

VENT

LARGE INTESTINE

SMALL INTESTINE

Chickens need proper nutritional support to maintain healthy laying and general wellness.

## Cleaning the Coop

How often you need to clean your coop depends on how many chickens you have, the size of your coop, the design of your coop, and the type of bedding you are using. If you have a coop with a poop tray of some sort, you may find that dumping it out every few days and scraping your roosts once a week will suffice in keeping smells and fecal buildup under control. If you have a coop that doesn't allow for quick cleanup, you might need to be more diligent to prevent ammonia from creeping in. Once you smell ammonia, you must clean your coop immediately. In general, weekly upkeep with a monthly deep cleaning is a good rule of thumb to keep chickens healthy and happy.

Every day, it's important to make sure the nesting boxes are free of feces or broken eggs and to refresh the bedding in the boxes if needed. If you have sand in your coop, you can use a kitty litter scooper with slots to remove fecal matter daily. A scoop like this can also come in handy if you are using shavings and want to remove visible poop daily or weekly. Since coops tend to accumulate dust quickly, you may want to keep a feather or synthetic duster in your coop for weekly dusting and cobweb removal.

For those times when deep cleaning is in order, scrub your roosts, nesting boxes, and other surfaces with one part white vinegar and one part water after they've been scraped of debris. Remove

## Homemade Parasite Repellent Spray

### WHAT YOU NEED

* 1 quart-size spray bottle filled with water
* 1 tablespoon dishwashing liquid
* 4 drops clove essential oil
* 4 drops peppermint essential oil
* 4 drops thyme essential oil
* 4 drops lavender essential oil

### DIRECTIONS

Add dish liquid and essential oils to the bottle and shake well. Spray on roosts once a week (don't spray it directly on your chickens). Shake well before each use.

---

### Aromatic Herbs for Birds!

I don't consider myself much of a gardener, but I love growing aromatic herbs for my chickens and either using them fresh or drying them to be used all year. When sprinkled about the coop, nesting boxes, and dust-bathing areas, they create a calm and fresh environment for my chickens (and me), while also helping to repel external parasites and flies. Some aromatic herbs that are great for chickens and easy to grow are lavender, mint, lemon verbena, and chamomile.

---

and replace all the bedding in your nesting boxes and coop with fresh material, thoroughly dust and remove all cobwebs, and wipe down all your windows both inside and out. If you've had an issue with mites or lice, consider spraying your roosts with a natural parasite repellent, which you can purchase or make at home (see above).

You don't need to hose down your entire coop to clean it properly, but as long as you allow it to thoroughly dry before roosting time, there's certainly no harm in doing it. One benefit of a plastic coop, such as the newer brands coming from the UK, is that it will dry quickly after being hosed down.

# Adult Chicken–Care Checklist

## EVERY DAY

○ Let them out at dawn or soon after.

○ Make sure waterers and feeders are clean and refreshed.

○ Remove fecal matter or other debris from nesting boxes.

○ Supply them with the correct feed for their life stage.

○ Offer a calcium supplement and grit separate from feed.

○ Observe chickens for abnormalities in behavior or appearance.

○ Spot clean as necessary.

○ Close them up at dusk.

○ Optional: Add immune-boosting herbs to feed (see Chapter 8).

## WEEKLY

○ Remove fecal matter from roosts and nesting boxes.

○ Replace nesting box material as needed.

○ Clean nipple and gravity waterers.

○ Refresh dust baths as needed.

○ Optional: Give them natural, weekly parasite prevention (see Chapter 8).

## MONTHLY

○ Deep clean the coop.

○ If possible, check over each chicken for external parasites or other abnormalities.

○ Optional: Give them natural, monthly parasite prevention (see Chapter 8).

## The Deep Litter Method in a Nutshell

If you're like me and would rather spend your time enjoying your chickens instead of shoveling their poop, the deep litter method might be for you. This method is said to have been developed in the late 1940s, though I wonder if it has some Indigenous roots. Basically, instead of frequently cleaning out your chickens' bedding and replacing it with new bedding, you add new bedding on top of the old in such a way that you transform the manure into beneficial bacteria-rich compost. When done correctly, the deep litter method can actually promote the health of your chickens, keep them warmer in the winter, and allow you to grow fantastically healthy vegetable gardens with the resulting compost.

1. Put about 3 inches of bedding on the floor of a clean coop.

2. When the bedding gets soiled, put another thin layer of bedding on top of it (usually about once or twice a week).

3. Rake the new bedding in with the old or allow the flock to turn it for you. If the bedding gets compacted, rake it to make sure it stays nice and fluffy.

4. In the fall or in the spring of the following year, remove most of the bedding, leaving behind a sparse, thin layer. Add the removed bedding to your compost pile.

5. Add fresh bedding, and rake it in the remaining litter to start the process over!

As you can see, the deep litter method is super easy. Just be sure that you don't add diatomaceous earth, a popular pest-control measure that we will discuss in Chapter 8, to your bedding because it may disturb the growth of beneficial bacteria. I have also found that deep litter works best in a dry coop. If you add ducks to your flock or you are having trouble keeping the moisture level down in your coop, this method may not be worth the hassle, at least at the beginning of your backyard chicken journey.

STEP 1

STEP 2

STEP 3

STEP 4

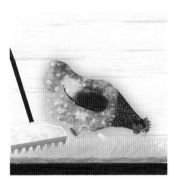

STEP 5

## Checking In with Your Chickens

Unfortunately, if chickens sense that a member of their flock is unwell or injured, they may bully or even attack that bird. Although it seems cruel, in their minds a weak bird makes the whole flock vulnerable to predation and so must be eliminated. Because of this aspect of the pecking order instinct, which is crucial for the survival of a wild flock but not desirable in a domestic situation, chickens are masters at hiding physical problems or

Observe each individual chicken's behavior every day, and check them over periodically for any abnormalities.

illness from their companions and caretakers. Every day, it's important that you observe your chickens' behaviors and appearance so that you can stay on top of any issues that might arise. If you notice that one of your birds is looking droopy or lethargic or is limping, or if something just appears "off," keep an eye on that bird and take appropriate action as needed.

At least once a month, do a thorough check of each chicken's condition so that you can catch any potential problems early. This might be a tall order for someone who has 50 chickens, but if your flock is small and you have a headlamp, it's a surprisingly easy task. Once your chickens are on the roost for the night and can be

more easily handled, you can head to the coop with your headlamp and check each one for lice, mites, or other abnormalities. If you have a larger flock and feel overwhelmed by this task, don't worry. Daily observation and a keen eye will suffice.

## When You Go on Vacation

Even the most loving and attentive chicken parent needs a vacation sometimes. If you've been able to set up your coop and run in a user-friendly fashion, your chickens shouldn't be too disrupted by having a new caretaker for a few days or even a few weeks. Your coop should have the desired characteristics discussed in Chapter 4 coupled with a simple food and water system that is easy for your chicken-sitter to follow.

If you have a predator-proof run, it may even be possible to leave your chickens unattended for the weekend without hiring anyone to watch them (though I recommend having a friendly neighbor drop by at least once to make sure everything is going smoothly).

If you are feeding your chickens soaked or fermented feed, it's okay to switch them to a pellet feed while you are gone. I recommend making that switch a week before you leave so you can be sure they are eating normally.

## Chicken-Sitter Checklist

○ Open coop at ＿＿ a.m. daily.

○ Collect eggs daily.

○ Refresh food and water every ＿＿ days.

○ Empty poop tray every ＿＿ days.

○ Close coop door at dusk daily.

○ Observe each day for abnormalities.

○ Phone number for emergencies and questions:

＿＿＿＿＿＿＿＿＿＿＿＿＿

# The Last Day of Spring

It was dawn on June 18, 1954. Just west of the heart of Guatemala City, in an area now called Zona 7, the community was already awake and bustling with activity. Through the narrow streets, buses stopped and started, edging around men pulling vegetable carts with two parallel handles and a single strap around their foreheads. Indigenous women, some with babies wrapped on their backs, hurried by impossibly fast, huge baskets of produce balanced on their heads.

A cheerful, fresh-faced 11-year-old girl meandered past a confused herd of cows emerging from an alleyway. Most people were on their way to the market, but she was headed to the bakery to buy traditional sweet bread for her family's breakfast. Though school didn't start for hours, the girl proudly donned her uniform: a button-down white shirt under a navy jumper. The songs of

numerous roosters echoed over the commotion.

A few moments later, sunbeams dotted the way as the girl ventured back to her family's tenement, breakfast in hand. In the faraway distance, she noticed Volcán Pacaya was sparking and smoldering, but this was a normal occurrence and no cause for alarm. The girl was happy and carefree, unaware that Guatemala was near the end of an era later dubbed The Ten Years of Spring. During this brief period, many common people, especially children, enjoyed the true promises of democracy. To the young girl, this meant education, health care, and the security of regular meals.

But it was also during this time that the Guatemalan government decided to relinquish some unused land owned by powerful foreign interests back to the Guatemalan people. One of the companies that lost land was the United

Fruit Company, which you probably know today as Chiquita Banana.

As the girl and her family sat down to dip their sweet bread into steaming mugs of hot coffee, an unfamiliar sound rang in the distance. The girl's siblings, two younger sisters and a baby brother, glanced nervously at her for answers, but she was just a child herself and couldn't understand what she was hearing. As panicked voices arose in the streets just outside their apartment door, the children's mother and father slowly came to realize what the sound was. It was machine-gun fire, and it was literally falling from the sky.

The girl in this story was my mother, and the event that took place that day, which sparked the devastating Guatemalan Civil War, was a coup d'état orchestrated by the CIA at the behest of the United Fruit Company. Of course, this event in history has often been framed in a different way, especially at the time. People said it was about Eisenhower's "hard line" against communism. They said it was about bringing freedom to the Guatemalan people. But those who lived through its atrocities know what it was really about: money, food, land, and power.

Though industrialization had already pushed my family out of their villages, I believe this event to be the defining moment that separated us from the land we walk on, the food we eat, the culture we experience, and ultimately the very essence of who we are. I don't think my family's experience is unique. If we all looked back into our ancestry, most of us would find similar happenings. Humans, unfortunately, have a tendency to focus on how different we are from each other, and when we suffer the consequences of those differences, we tend to blame each other. But what if, by finding the true culprits of our separation, we found that we were all on some level suffering from the same affliction?

Chickenlandia has been a godsend in my life in terms of slowly finding my way back to what was taken from me. Each moment I spend in the chicken yard is an opportunity to awaken my sense that the natural world is part of who I am. My hope is that the more each of us has a similar opportunity, the less time we will spend blaming each other for the loss of our Ten Years of Spring that we don't really remember but we know was once there.

## Let's Unlearn
# CHICKEN NUTRITION

There are numerous valid and sustainable ways to feed chickens, including feeding them only kitchen scraps and allowing them to forage for the rest of their diet, but this isn't optimal for all chickens in all cases. As much as I wish things were different, I must acquiesce that chicken feed, fortified with certain nutrients to support modern chickens, is important in a flock's diet. But I refuse to accept that meals of solely processed feed are preferable to ones that also include fresh nutrients from healthy kitchen scraps, foraging or fodder, sprouts, and other nutritious treats.

## Chickens Can Eat People Food

Have you ever wondered why there is chicken feed? I know this might seem like a strange question, but if you think about it, chickens went without commercial feed for thousands of years—for most of their existence on earth. In fact, many people throughout the world still don't give their chickens commercial feed, even though there is a strong push to "educate" people (most of them poor) about its "necessity."

Now, I want to make it clear off the bat that I'm not advocating that you shun commercial chicken feed altogether. This is a beginner's book, and I believe it's best to lay out reasonable, accessible steps for people from all walks of life in our quest for sustainability. It's also an unfortunate truth that some chickens are better able to tolerate a scavenged diet than others. Many breeds from just a few generations

As long as they've been domesticated, chickens have been fed scraps and wasted food.

ago were biologically different from the ones raised today. Modern laying hens have been bred to produce an incredible number of eggs compared to their ancestors, and as such have specific nutritional needs that are best met through commercial feed. In the absence of layer feed, your chickens may not lay as well and possibly won't live as long as they would if you added it in. Still, the idea that in 2018 alone the world produced 1,103 million metric tons of chicken feed has me wondering, When and why did the way we care for chickens stray so far from its sustainable roots?

When I was in my early twenties, I found myself working at a pet food and supply store in Arizona that was owned by two women who were impressive pioneers in the holistic animal care field, although

I didn't realize it at the time. The store sold dog, cat, and livestock feed, as well as homeopathic and herbal formulas tailored for creatures great and small. One of the more controversial ideas held by these women was that cats, dogs, and other animals should be given fresh food that was appropriate for their species in addition to a healthy feed. In other words, we should give our fur and feather babies people food. *Gasp!*

I was shocked when I first heard their thoughts on this. Wasn't "people food" bad for pets? They assured me that pet and livestock feed was a relatively recent invention, and that I only thought the way I did because it had been cleverly drilled into me by the marketing companies of large corporations. After all, most commercial feed must undergo high temperatures during processing, which causes it to lose nutrients. Vitamins and minerals must then be added back into the feed to make it nutritionally complete. "How well could you live on dry, processed pellets?" they asked. A light bulb went off inside me. What these women were saying, no doubt often in the face of loud opposition, made perfect sense. I knew they were right.

## Chickenlandia's Chicken Food Pyramid

I'm all about commonsense, uncomplicated practices that everyone can follow and adopt as second nature. This is why I created the Chicken Food Pyramid, a basic plan to help you deliver a balanced, healthy diet to your chickens without getting consumed by measurements or overly strict guidelines. While you could make your own nutritionally whole homemade feed, I geared this process toward beginners or just busy people for whom it makes the most sense to give their flocks a diet of mostly commercial feed.

**The first tier** is chicken feed that is appropriate for the age of your flock. It's the largest segment since it's the foundation of their diet.

**The second tier** is healthy kitchen scraps. Focus on vegetables (leafy greens are best) and low-sugar fruit. I know they aren't necessarily leftovers, but you can also add sprouts to this level; we'll learn how to grow those later in this chapter.

**The top tier** is healthy treats. This is where purchased items such as mealworms, grubs, and scratch grains (a combination of grains but not a complete feed) fall. Homemade treats will also fall here, as well as proteins and higher-calorie scraps like scrambled eggs, whole milk yogurt, leftover corn, and shrimp tails.

**HEALTHY TREATS**

**HEALTHY KITCHEN SCRAPS**
Mostly green vegetables with other veggies and low-sugar fruits

**LAYER FEED**
Needed to meet the nutritional requirements of laying hens

## A Note on Fatty Liver Hemorrhagic Syndrome

One of the main concerns that some chicken educators and veterinary professionals have about feeding chickens anything other than their chicken feed is that they could develop a condition called fatty liver hemorrhagic syndrome (FLHS), which can lead to sudden death. Obviously, this is a tragic occurrence, and we want to avoid this if possible. However, I think the wrong culprit is often blamed. Let me explain.

FLHS is commonly said to be caused by too many "high energy" foods and a lack of exercise. To put it simply, many vets and educators will tell backyard chicken owners that they overfed their chickens and that's why they died. But how can this happen to a flock that lives in a backyard where they're fed healthy scraps and appropriate feed and have room to peck, scratch, and run?

In my opinion, this is one of those instances where we've been misled by scientific studies of factory-farmed chickens. Of course, layers in factory conditions are developing FLHS from overfeeding and lack of exercise; they live in tiny cages where they can't even move and the only thing they have to do is eat! But FLHS, in both industrial farming and backyard flocks, can also be caused by hormone fluctuations in high-production hens, mainly during the months when they are laying the most eggs.

This leads me to what I believe is a reasonable assumption: When well-cared-for backyard chickens develop FHLS, it has more to do with being bred to produce an unnatural number of eggs than it does their healthy, balanced diet. We need more studies of backyard flocks to prove this is true, but I don't rush to blame healthy scrap feeding as the reason for FLHS. After all, domestic chickens are scavengers by nature. Eating our healthy food scraps is something they were born to do.

## The Benefits of Fermented Feed

The simple practice of fermenting feed not only saves money but also has many benefits for chickens, such as boosting their gut health and immunity to disease. It is one of the best things I ever learned for the sake of my family and flock.

I use lactic acid fermentation, which people have used for centuries to preserve food while also making it more digestible and its nutrients more bioavailable. Lactic acid fermentation relies on *Lactobacilli* (naturally occurring bacteria in our environment), which convert sugars (carbohydrates) in food into acid. All you need to do is add water and wait for the bacteria to do their thing. What you end up with is a bubbly, tangy slurry full of friendly bacteria that, when given to your flock, helps to create a healthier gut, crop, and vent, while also benefiting their immune systems. I normally ferment a raw mash feed, but you can ferment pellets, crumble, non-medicated starter, grower feed, and even scratch (as a nutritious treat).

When chicken feed ferments, it expands to about twice its original volume. Since the nutrients in the feed also become more bioavailable, your flock may actually end up eating less of it each day. For standard-size birds, I suggest starting them out on ¼ cup of fermented feed daily and then adding more as needed. For bantams, try ⅛ cup. Your chickens will likely need more than these initial servings, but starting with a smaller portion size will help you to avoid waste and get the most accurate idea of what your flock will consume by nightfall. You definitely don't want to leave fermented feed out overnight, even though the beneficial bacteria are wonderful for rodents as well!

If it just doesn't make practical sense for you to ferment your chickens' feed daily, you can always offer it on occasion. Give it as a treat or when you sense they need some extra support, such as when there is a sick chicken in the flock or during parasite infestation.

---

### Creating a Fermentation System

It's best to make your ferments in batches that will be entirely consumed in a day. To achieve this, I suggest creating a system so that when one jar is ready, there are two more still in progress.

Each day, feed one jar to your chickens and start fermenting a new jar. If you do end up with some leftover feed, you can store it in the fridge for up to 2 weeks, then bring it back to room temperature when you are ready to give it to your flock.

---

# How to Ferment Chicken Feed

I recommend you start small by fermenting 1 cup of feed, and then make larger batches as you become more confident with the routine. I specify 2 parts water for 1 part feed, but feel free to play around with these measurements until you reach the consistency you want. If you find you want a drier end product, start out with a little less water. Or add more water for a soupier result. Raw mash feeds tend to ferment the best, but plenty of people ferment crumble and pellets with no problem, though I have found there to be some variation among different brands.

Note: If your ferment smells "off," has developed mold, or smells alcoholic, *never* feed it to your chickens, and especially not to your baby chicks.

## WHAT YOU NEED

- ❖ 1 cup pellets, crumble, mash, non-medicated starter, or scratch
- ❖ 1 quart-size jar or plastic container with lid
- ❖ 1 spoon
- ❖ 2 cups non-chlorinated, well, or distilled water

## DIRECTIONS

1. Place the feed in the jar.

2. Pour the water over the feed.

3. Stir and cover loosely with the lid so any gasses can escape.

4. Store your jar at room temperature out of direct sunlight for up to 3 days, stirring at least once a day. When your ferment is bubbly and smells tangy like yogurt or sourdough, it's ready to feed to your chickens!

# How to Sprout Grains and Seeds

There are a few different sprouting methods, but my preferred technique for a small backyard flock uses a quart-size jar. You can sprout numerous grains, seeds, and legumes for your chickens, and I've listed some of my favorites below. Just keep in mind that this method won't work for gelatinous seeds like flax or chia.

## WHAT YOU NEED

❖ 1 quart-size jar with lid
❖ 1 strainer lid or cheesecloth with rubber band to secure it
❖ 3 tablespoons grains or seeds (my favorites: mung beans, sunflower seeds, lentils, broccoli seeds, alfalfa, barley, wheat, kohlrabi, garbanzo beans, and clover)
❖ Water

## DIRECTIONS

1. Place the grains or seeds in the jar and put the strainer lid on or cover the jar tightly with cheesecloth. Rinse the contents by filling the jar with water through the lid and letting it drain out several times. Remove strainer lid and pour in fresh water to cover the grains or seeds by about 3 inches. Put the regular lid on tightly.

2. Let the grains or seeds soak, covered, at room temperature for up to 24 hours or until the grains or seeds have opened a little.

3. Replace the regular lid with the strainer lid or cheesecloth and drain the sprouts. Rinse the sprouts well, then set the jar either on its side or in a sprouting stand (a special stand made for sprouting grains and seeds in a jar) in a sunny spot at room temperature. Thoroughly rinse and drain the sprouts twice a day. You want to keep them wet so they can grow, but you also don't want mold to develop.

   You can feed these to your chickens at any point after they start sprouting, but I like mine to be slightly green, which usually takes about three or four days. After they are about 2 inches long, if you're not going to feed the sprouts to your chickens right away, you can store them in the refrigerator.

## The Benefits of Sprouted Grains and Seeds

I know that many of you are working with limited space, but it's a hard fact that pasture-raised chickens are generally healthier and lay healthier eggs than chickens that do not have access to diverse greenery, rich soil, and bugs. If you must keep your chickens in an enclosed run, do not despair! Sprouting grains and seeds is one simple way to get fresh nutrients typically found in the pasture into your flock. Like fermenting feed, this process is surprisingly simple and fun, especially for kids.

I don't strictly limit the amount of sprouts I give my flock because they are just so healthy for chickens, falling in the second tier of my food pyramid. Your chickens may devour them, or they may be suspicious of them at first. I would not let that discourage you. Eventually, they will learn how wonderful sprouts are! Start by giving them small amounts and experiment with different seeds and grains. Remember: If you sprout something that your chickens don't like, you can always eat the rest yourself! *Yum.*

## Pasture-Raised in Small Spaces

When I put my first flock outside in their enclosed run, I was shocked at how quickly they devoured and destroyed every last morsel of plant life within their reach. As you probably know by now, chickens will make short work of most greenery available to them (some plants they won't eat), unless they have a very large area in which to graze. But what if I told you there was a way that your chickens could get some good nutrition from fresh fodder grown right in their chicken yard (with the bonus of producing awesomely orange yolks!) and that you could create this system easily, regardless of space? Enter the magic of chicken salad bars.

No, I'm not talking about chicken salad that you put on a sandwich! I'm talking about something you can build yourself and place in your chicken yard that will allow plants to sprout protected from your chickens, but then allow your chickens to eat the

A chicken salad bar is easy to make and can offer your flock needed nutrition they may not receive without access to pasture.

plants when the plants are tall enough. One way to do it is to make a frame by nailing or screwing together two-by-fours in a square or rectangle, then stapling hardwire mesh on top. You can also use chicken wire or netting, though it won't be as sturdy. Plant seeds in the dirt of your chicken yard, then place the salad bar directly on top. The plants then grow up through the wiring or netting without giving the chickens a chance to destroy them before their sprouts even make it above ground.

Another way to create a salad bar is to plant seeds in a wooden planter, then staple hardwire mesh or chicken wire over it (you may need to remove and re-staple when it's time to reseed). You don't even need to build anything or buy new materials—it's a cinch to use leftover hardware cloth, chicken wire, or netting to shield seedlings.

Grasses like wheat and barley are perfect for salad bars because they grow well through mesh, wiring, or netting. You can also buy packaged seed medleys that are made just for chickens, which usually contain seeds for growing flax, clover, alfalfa, and other nutrient-rich plants. You may have noticed that pasture and salad bars don't appear on my Chicken Food Pyramid. This is because I simply do not believe in measuring them. Chickens should have access to foliage at all times, without limit. Not only does it provide much-needed nutrients, but in my opinion it is essential to their experience here on Earth.

## Supplying Calcium

Once hens reach laying age, you will need to add calcium to their diet. This essential mineral helps them develop strong bones and eggshells, along with other benefits. The most touted calcium supplement among chicken enthusiasts in the United States is crushed oyster shells, which you can purchase online or at most farm stores. Limestone packaged for chickens is another option, as well as dried soldier fly larvae, although the larvae are more expensive.

A sustainable and age-old calcium supplement would be to crush and dry your flock's own eggshells and then feed them to your chickens. Although some say this is an inadequate source of calcium, in the

Offering your chickens calcium-rich food scraps and their own crushed eggshells can aid in healthy laying.

## Calcium-Rich Vegetables

- ◯ Arugula
- ◯ Beet greens
- ◯ Bok choy
- ◯ Broccoli
- ◯ Collard greens
- ◯ Kale
- ◯ Mustard greens
- ◯ Okra
- ◯ Spinach
- ◯ Swiss chard
- ◯ Turnip greens
- ◯ Watercress

interest of inclusivity and with the knowledge that this has been practiced for quite some time, I mention it as an acceptable alternative. You will want to be mindful about offering plenty of calcium-rich vegetables, either whole or as table scraps, to your chickens. (Hint: These veggies are all great for your family as well!)

## Don't Forget the Treats

I hope by now I've clearly communicated how strongly I feel that keeping chickens is important not just for sustainability and humans' physical health, but also for how it encourages a connection with nature and, consequently, our shared ancestry. When I'm outside with my flock, tossing them scratch, mealworms, or scraps, I feel a moment of peace in my otherwise hectic life that is often disconnected from natural rhythms. These moments are important for my well-being, possibly in ways I'm not even aware of.

All the above is to say that I believe in feeding chickens treats. Of course, you don't want to overdo it, but if you make treat-giving a daily ritual, it will bring you so much fun, connection, and possibly healing that I think it's good not only for your chickens but for you as well. You can buy packaged treats, such as mealworms, grubs, scratch, or cracked corn, online or at your local farm store. But you can also treat your chickens with numerous things that you likely already have in your kitchen or that you will have left over from your next family meal, such as:

- ✤ scrambled eggs
- ✤ cooked or raw corn
- ✤ peas
- ✤ strawberry tops
- ✤ cultured cottage cheese
- ✤ cooked grains
- ✤ shrimp tails
- ✤ canned tuna (no salt)
- ✤ cooked unprocessed meats
- ✤ whole milk yogurt

When I'm feeding treats, I'm mindful about where they fall in the Chicken Food Pyramid. I don't want my chickens to eat so many treats that they aren't getting the appropriate amount of feed or greens. But I will never be that person who says treats are bad. Everything listed on page 111 has nutritional value and will add to the diversity of vitamins, minerals, and other nutrients your flock needs. So please, take a breather, sit down among your flock, and feed them by hand. Let that be your meditation and your moment to connect with something that feels rare in our busy world.

When I'm outside with my chickens, tossing them something tasty, I aim to feel

the same way my ancestors did: like I'm doing something that just makes sense. A bonus is that I've become more mindful about what I feed myself and my family because repurposing resources is so much better than throwing them away.

---

## The Feeding Routine of a Healthy Flock

**Morning:** Top off their feed containers or give them their daily ration of fermented feed. Toss them a handful of grubs, mixed veggies, or other appropriate treats.

**Afternoon:** By this time they should be finished with their ferment and are pecking and scratching for insects or having a bite at their salad bars. You can throw them some healthy leftover scraps from lunch with some homegrown sprouts mixed in.

**Evening:** If their day didn't include many scraps or if it is especially cold, you can throw them a warming treat such as cracked corn or maybe a combination of leftover grains with mealworms or other lean protein.

Spend none of your day on meticulous measurements. Know that your chickens are well-nourished, healthy, and happy.

# PREVENTING DISEASE
## Naturally

Normally, your chickens' natural habitat should not make them sick. Your flock should be able to take advantage of what's beneficial around them while not becoming overwhelmed by what's harmful. But how do we make sure our chickens stay healthy and are not overcome by the harmful microorganisms around them? Some factors are beyond our control, but one is not: stress. Stressed-out chickens are much more vulnerable to all kinds of problems, including sickness and parasitic infection—what I refer to as "dis-ease." When chickens are not at ease with their life, they are more vulnerable to all kinds of issues, health and otherwise.

## There's More to Fresh Air Than Oxygen

As I write this page, I'm sitting on my 7-year-old's bed, keeping him company as he drifts off to sleep. It's May in the Pacific Northwest, and there's a gentle spring rain outside. I have the window open, and the fresh air is drifting in. It's especially nice tonight because along with the normal Northwest aroma, there's a bouquet of wet evergreens. I've been trying to write for hours and have had such difficulty getting words on the page. But now I feel peaceful, and my creativity is flowing. Is it weird that I feel my change in attitude might be connected to the open window? Could it be that there really is "something in the air"?

Today's world is strange. On one hand, we are taught that we must "get outside" to be healthy, usually in the context of staying active. On the other hand, we are taught to fear the outside. We're constantly warned about harmful germs lurking in the dirt, in the air we breathe, and in the water around us. But we don't really discuss how we also need germs to survive. In fact, life on this planet could not exist without bacteria, protozoa, fungi, and all the other characters of the micro-world. Just think of how all the creatures we *can* see are essential to each other, then consider that two-thirds of life on this planet is made up of organisms we *can't* see. Even viruses, often thought of as purely adversarial, are part of the building blocks of life. As we discussed in Chapter 2, the role these things play in our existence is likely more important than we know. And yes, they play a critical role in our chickens' lives as well.

Now, I don't want to mislead you by suggesting there isn't some scary stuff out there. There are, unfortunately, devastating and aggressive diseases that could seriously harm your flock. To mitigate these dangers, I've seen all kinds of suggestions, from keeping baby chicks inside for an extended period to completely enclosing your chickens so that no wild bird ever flutters near them. One of the reasons why chickens are kept indoors in factory settings is that it's easier to control their environment and limit exposure to pathogens. I understand these sentiments and acquiesce that in these modern times perhaps there are situations when such drastic measures are necessary. But ultimately, I think domestic chickens belong outside, participating in the world they have occupied for thousands of years. I believe they actually need constant exposure to that "something in the air," not just for their physical well-being but for their mental well-being as well.

## Four Easy Steps to Avoid Disease in Your Flock

1. **Provide early and unlimited exposure to nature.** Baby chicks and chickens belong outside. Offer your baby chicks gentle, controlled exposure to their outdoor environment, and try not to limit your flock's interaction with nature throughout their lives.

2. **Support your chickens with good nutrition.** Fermented feed, if you can make time for it, is a wonderful way to provide a solid nutritional foundation for your flock. If you can't ferment, that's okay. Just make sure your flock has the appropriate feed for their age and give them lots of nutrient-dense scraps, sprouts, and access to pasture or salad bars.

3. **Make sure your chickens have enough space.** They also need enough enrichment in said space. Crowded chickens are stressed-out chickens. Bored chickens are stressed-out chickens. Stress is a harbinger of disease. Not to mention that crowded spaces are often dirtier than they should be.

4. **Do your best to practice good husbandry.** This is where having a user-friendly coop, as discussed in Chapter 3, is so important. A coop that is not properly maintained may harbor disease-causing ammonia fumes and feces-loving pathogens, and hastily spread any illnesses that may pop up.

## Parasites Need Love, Too

Oh, parasites. You may itch and squirm simply at their mention. I understand this. Parasitic infestations are certainly awful to deal with, and they are one of the main reasons why people decide to throw in the chicken-keeping towel altogether. I'm hoping that I might (mite?) be able to help you feel more at ease about these creepy-crawlies with a little education about them. Perhaps knowing more about their purpose will soften the blow should they decide to occupy your flock's space.

I must confess that while most people find parasites disgusting, I find them fascinating. Did you know that there is a nematomorpha (parasitic hairworm) in Japan that upon infesting crickets and grasshoppers, makes them want to drink more water? This might not seem like a big deal, but when the thirsty crickets and grasshoppers travel to the river for a drink, they become a main food source for an endangered trout. Can you now see how this pesky parasite is a critical part of the ecosystem? And they are not alone. Parasites are an indispensable contributor to the biodiversity of our planet. Their presence or absence can affect how likely we are to be exposed to certain diseases, how our food grows or doesn't grow, and even how our oceans survive. Yeah, they're *that* important.

Even so, right about now you're probably thinking, "This is all well and good, but I really don't see how mites, lice, or roundworms are actually beneficial to my

Parasites play an important role in controlling populations of wild animals.

chickens." I can certainly see your point, so let's go deeper.

Your flock can be exposed to parasites in numerous ways, but one common source of transmission is through wild birds. One important role that parasites perform is to weed out the sick or weak. This controls the size of the bird population and strengthens the species. Parasites have evolved alongside their hosts for thousands of years to serve this important purpose. If your chickens were living 8,000 years ago in the jungles of Asia, parasites would be critical to their survival. Of course, your flock is not living in jungles past, so it's completely understandable that you don't want them to die from anemia or some other awful parasite-induced condition. Yet, if you change your outlook on these supposed freeloaders, you can see how their role is still important today, even in a domestic flock.

Unless they have been preemptively medicated, almost all backyard flocks have some parasites. This is a normal condition of a healthy flock. Issues arise when a bird has an infestation or infection, meaning that it has too many parasites for its system to overcome and it falls ill. But here's the thing: If a chicken has an infestation or infection, it's most likely there was already something going on that created the right conditions for them to become vulnerable.

Remember, part of a parasite's job is to get rid of sick or weak animals. So, instead of viewing them as the enemy, imagine the little worms or mites or protozoa waving tiny flags and shouting, "Hey! Something is out of whack in this ecosystem, and you need to fix it!" This is your opportunity to restore balance in your flock before the parasites do it for you.

If you should find yourself observing tiny critters crawling all over your favorite chicken or staring in shock at a worm-filled poop (yuck), the very first thing I want you to do is evaluate your practices. Are your chickens getting the right nutrition? Do they have enough space? Are your husbandry habits where they should be? Like I said before, these practices are so simple, but they are your flock's foundation and can profoundly affect their physical or mental health. Even if your chickens seem fine physically, stress itself can make them vulnerable to parasite overgrowth. It's all part of their little mini-ecosystem in which you are the stand-in for Mother Nature.

Having said all that, please don't beat yourself up should parasites become an issue in your flock. Sometimes, problems simply cannot be avoided, especially if you adopt birds from other places and/or if your flock lives on a smaller piece of land. My flock has dealt with lice, roundworms, and the bane of my existence: scaly leg mites. This is to be expected when you rescue chickens as I often do. But my greatest task when I bring home new chickens is to restore balance. Nature isn't personal. It just gives us information so that we can return to harmony with it.

## Building Resilience from the Inside and Outside

I cannot stress enough the importance of healthy gut bacteria for your chickens, especially if they are going to be living on a smaller lot where they won't have as much access to the beneficial bacteria in the soil as pastured chickens do. When your chickens have a healthy gut, they can better fight off disease or an infestation of parasites.

We know that mother hens encourage their baby chicks to explore their natural environment soon after they hatch. We also know that however clean and well-kept a chicken yard is, there will be microbes in the coop and in the soil. This is completely normal. Healthy chickens who have lived in the same space for a while will normally develop immunity to the germs around them, and baby chicks can usually develop that same immunity with slow exposure in a low-stress situation, like we discussed in Chapter 5.

Early exposure to the outdoors helps baby chicks build immunity to microbes in their natural environment.

In addition to fermenting their feed, which offers them wonderful support for their digestion and immunity, there are a few other immune-boosting and health-supporting items you can offer your flock from chick to adulthood, including water with electrolytes, vitamins, and probiotics; apple cider vinegar; herbs; and garlic. These supplements can help prevent disease and parasite infestation throughout their lives. Of course, no herb, tonic, or even synthetic medicine is foolproof.

You could meticulously follow every step in this book and still lose a chicken or chickens to illness. The fact of the matter is, sickness happens. But as we know with our own bodies, it's important to try to give our flocks their best shot at wellness by being mindful of their care, optimally from the get-go. Of course, if you already have chicks or adult chickens, rest assured it's never too late to start with some natural wellness practices.

## ELECTROLYTES, VITAMINS, AND PROBIOTICS

An electrolyte, vitamin, and probiotic (EVP) powder, when mixed with water, will give your chicks a nice boost of vitamins, extra hydration, and beneficial bacteria that will help to prevent issues such as starve-out and pasty butt. You can find the powder at your local farm store, online, or at large farm store chains. When your chicks arrive home, offer them tepid water with the appropriate dose of EVP (follow the instructions on the label). The taste should encourage them to drink it.

After one week, if your younglings are lively and growing well, you can discontinue the EVP use. If you have any chicks that seem to not be thriving, you can continue with it for another week. As an alternative to the powder, you can make your own electrolyte water (see sidebar at right).

## APPLE CIDER VINEGAR

Apple cider vinegar (ACV) is also full of benefits for your chicks, and you likely already have it in your cupboard. Just be sure it's the kind that says "with the mother" so that you know it's full of beneficial bacteria, plus vitamins and minerals

# How to Make Homemade Electrolyte Powder

If you don't have access to store-bought EVP powder, or you would just rather make something homemade, try this simple recipe for baby chicks and chickens. Using raw honey will add some vitamins and other nutrients, including prebiotics, which are good for gut health.

### WHAT YOU NEED

* 4 tablespoons baking soda
* 4 tablespoons sea salt
* Raw honey

### DIRECTIONS

1. **To make the electrolyte powder,** mix the baking soda and salt together, then store it in a cool, dry place.

2. **To make 1 gallon of electrolyte water,** add 2 teaspoons of your electrolyte powder and 1 tablespoon of honey to 1 gallon of water and stir.

3. **To make a slurry for sick chicks,** add a pinch of electrolyte powder and ¼ teaspoon of honey to 1 cup of cooled green tea and stir. (See How to Make the Sick Chick Slurry, page 134.)

that are great for growing biddies. One very important benefit of ACV is that it is known to ward off disease, namely coccidiosis, which we will be discussing more in the next chapter.

You can add ACV to your chicks' water (1 teaspoon per quart) when you first bring your chicks home, but if you're starting with EVP, finish their course of that first and then begin with the ACV. With a few exceptions, chickens will benefit from ACV for their whole lives.

## IMMUNE-BOOSTING HERBS AND GARLIC

Numerous medicinal herbs have been observed to be valuable for both baby chicks and adult chickens. The top two that I suggest, both for their accessibility and their health benefits, are oregano and thyme. You can purchase a prepackaged herbal supplement specifically for chickens that contains these two herbs as well as others, but you can also grow them yourself rather easily. I like to sprinkle dried oregano and thyme around the brooder when the chicks first come home. Later, I give them fresh stems and leaves to play with and eat. Sprinkle these herbs in your brooder or chicken yard a few times a week or even daily when your birds need a boost, such as during illness or stress. The herbs will benefit your chickens for the entirety of their lives.

Garlic is often referred to as an herb, but it is actually a vegetable. With too

Oregano and thyme are easy to grow and can help boost your flock's immunity to disease.

many health benefits to list, it is well known for its immune-boosting and anti-parasitic abilities, including preventing coccidiosis. When your chicks are about 2 weeks old, you can begin adding one-quarter of a fresh clove per quart of water to your chicks' waterer. Make sure you watch your chicks closely to ensure they are drinking and not getting turned off by the taste or smell. You can also mince garlic and add it to their feed, or mix it in with a little hard-boiled or scrambled egg. You can supplement garlic three times a week or every day if there is illness or a reason for concern about illness in the flock. Once they reach adulthood, increase the dose to a full clove of fresh garlic per gallon of water or one finely minced clove with their feed.

## EXTENDED MEASURES FOR PARASITE PREVENTION

Regular use of the above measures, along with other good husbandry practices, is a wonderful way to keep down the parasite load in your flock. However, if your chickens live in an enclosed run for the entirety of their lives, or if your flock's environment has a particularly high parasitic presence for whatever reason, they may need additional protection. Depending on what country you live in, you can purchase herbal tonics, usually marketed as "digestive aids," that help with parasite resistance and work them into your chicken-care regimen. You can also find herbal formulas for treatment, which we will discuss further in the next chapter.

But there is one more thing I want you to be aware of. In many modern chicken-keeping circles, the above natural preventives are considered inadequate. Instead, the idea of preemptively deworming your flock, usually with synthetic medicines that may or may not be approved for use in laying hens, is aggressively promoted regardless of the environment in which the chickens live and whether or not there is an actual confirmed parasitic presence in the flock.

I don't believe that conventional dewormers are bad in and of themselves or even that using them off-label when necessary is dangerous. But if most backyard chicken keepers as well as industrial farmers administer them several times a year, year after year, doesn't it make sense that they would affect the environment?

Just like the overuse of antibiotics has caused bacteria to evolve resistance against them, parasites can also build resistance to medications designed to kill them. Furthermore, these medications could pass from chicken feces into the soil, possibly harming other parasitic and microbial life. Taking into consideration the interconnected role that parasites play in the cycle of life, the regular use of anti-parasitic pharmaceuticals could have serious negative consequences on the environment.

I know I probably sound like a broken record, but once again, I believe the focus should be on giving your flock everything they need nutritionally, physically, mentally, and emotionally. In my opinion, the only worthwhile preemptive measures are those that seek to protect the earth we all live on.

## Medicated vs. Non-medicated Feed

One of the most common microorganisms found in a chicken yard is a parasitic protozoan called coccidia. Most adult chickens

have one or more species of coccidia living within their bodies, but over time have acquired immunity to widespread infection by the pathogen. Baby chicks, when gradually exposed to their outside environment, by either a mama hen or a human, have a good chance of also acquiring this immunity. There are instances, however, when a chicken's digestive system becomes so overrun with coccidia that they develop a severe form of gut inflammation called coccidiosis. This usually happens when a chicken is compromised by another illness; lives in soiled, muddy, or stressful conditions; or is abruptly introduced to an unfamiliar species of coccidia. It is most common in baby chicks.

A chicken or baby chick infected with coccidiosis will appear lethargic, hunched over, and droopy, and it might even have blood in its stool. If one of your chickens is displaying these symptoms, their risk of death is high, and your flock is in grave danger. Coccidiosis is highly contagious, especially if your birds are living in damp or dirty conditions. I don't want to downplay the often-devastating consequence of this disease because it's a real problem and has rightfully earned a terrible reputation among chicken keepers. But I'm afraid in our fear of it, we've created a situation where our solution might be more damaging than the disease itself.

When you go to the store to buy baby chick starter, you will notice that you can choose between medicated and

Non-medicated starter feed is the best choice for small backyard flocks.

non-medicated feed. Medicated chick starter contains an ingredient called amprolium. Amprolium is the most widely used medication for an active case of coccidiosis, but it also serves as a preventive when given in a lower dose, as in medicated chick starter. Amprolium works by mimicking vitamin B1 (thiamine), which coccidia need in order to reproduce. When coccidia are tricked into consuming amprolium rather than thiamine, they cannot continue their life cycle and die off. Since the dosage of amprolium is low in medicated feed, enough coccidia will perish to prevent coccidiosis while still allowing the chick enough exposure to have a good chance at developing immunity.

When compared to numerous other medications, amprolium is fairly safe. It's been around for a long time and requires no egg withdrawal period when given to laying hens, meaning you and your family can continue to enjoy your flock's eggs during and after amprolium is administered without worry that the medication will transfer to you via those eggs. This does not, however, mean that either a preventive or curative dose is totally benign to your birds. Coccidia need thiamine, but so do baby chicks and adult chickens.

Thiamine is often referred to as the stress vitamin for its role in helping birds (and humans) deal with stressful situations, which we know is crucial when it comes to overall flock health. Adult chickens suffering from thiamine deficiency can pass that deficiency to their young, making them more vulnerable to stress, starve-out, lethargy, tremors, and even death. In fact, a common neurological condition in baby chicks called wry neck can be caused by a thiamine deficiency either from the time of hatch or due to the use of medicated feed, but we don't often hear about this as the cause.

Medicated feed does have one very strong benefit: It prevents coccidiosis most of the time. Its use is certainly necessary in factory farms where the conditions are ripe for infection. But is medicated feed really needed in small backyard flocks with good practices? As I mentioned before, coccidia need certain conditions to multiply to the point where their presence triggers coccidiosis, conditions that are hopefully not going to be found in your backyard. If you keep your chicken coop and yard reasonably clean and free of mud, make sure your flock is getting proper nutrition, and keep the stress in their life low, these basic practices will be your best insurance. Oregano, thyme, garlic, and apple cider vinegar are also wonderful at preventing this disease.

Of course, none of the things I mentioned, including giving chicks medicated feed, is an absolute guarantee that your flock will be coccidiosis-free. But when calculating risk, I believe that it is best to avoid medicated feed and instead focus on natural prevention and gradual immunity. In the next chapter, we'll talk about what you can do if you get an active case of coccidiosis in your flock.

## Mites, Lice, and Dirty Chickens

Chickens do not voluntarily bathe in water as various other bird species do. Instead, they find dry, dusty areas of earth in which to writhe around, loosening any foreign material as well as external parasites from their feathers and body. This activity is called dust bathing. Their flurry of wiggling and contorting is rather comical to watch, but it can also be panic inducing for new chicken keepers who mistakenly think their chickens are having a seizure! Not to worry, they're just taking a dust bath, and even baby chicks love to do it if given the opportunity.

It's very important to keep your chicken coop and yard reasonably clean and free of mud and feces buildup to thwart internal and external parasitic

Chickens clean their skin and feathers by bathing their bodies in dry sand or dirt.

infections. It's also equally critical for your chickens to have a place to dust bathe year-round. During the spring and summer months, it is usually easy for chickens to find dry areas in which to roll around, depending on the climate you live in and the terrain of your yard. But in the fall and winter, it becomes more difficult. During these months, you need to be sure to provide a dust-bathing spot in a location that is not exposed to the elements. For baby chicks, you can simply give them a pie dish or a tray with a little dry dirt. For adults, you may want to add in some additional parasite-repelling ingredients, like wood ash and diatomaceous earth.

## How to Make a Chicken Dust Bath

It may take your chickens a few days to figure out that they can bathe in their homemade dust bath, but they should figure it out eventually. And just so you know, all you really need for a decent dust bath is dry dirt or sand. The other ingredients provide extra protection against parasites but are not necessary.

### WHAT YOU NEED

- ✤ Large horse feeder pan, cat litter box, wooden planter, or other nontoxic container that a chicken can get into
- ✤ 4 parts dry dirt or sand
- ✤ 1 part wood ash or charcoal (optional)
- ✤ 1 part food-grade diatomaceous earth (optional)
- ✤ A couple handfuls of dried aromatic herbs such as mint, lavender, lemon verbena, chamomile, or eucalyptus (optional)

### DIRECTIONS

Place the dirt or sand in your container. Add the wood ash and food-grade diatomaceous earth, if using. Sprinkle in the aromatic herbs, if using. Place in a dry area, away from the elements. Watch your chickens bathe in it!

## A Word (or Two) about Diatomaceous Earth

Few substances elicit such passion (and scorn) among people who love chickens as diatomaceous earth. The passion arises from genuine concern for the safety of both humans and birds, but unfortunately, some well-respected and otherwise knowledgeable sources have given what I consider an alarmist misrepresentation of diatomaceous earth. I believe in making informed decisions with all the facts present, so let's break down the two kinds of diatomaceous earth and clarify which type I recommend using in the chicken yard.

Diatomaceous earth is a natural substance made from the skeletons of fossilized single-celled algae called diatoms. Diatoms have intricate cell walls of silica, giving the tiny aquatic creatures a glass-like appearance. When silica, a trace mineral we all need, is in its natural state and not affected by human-produced or volcanic heat, it is referred to as amorphous and is generally considered safe. Amorphous diatomaceous earth is found all over the planet and is used in lots of things we come in contact with daily, including makeup, toothpaste, and even food we eat. It is sometimes added to animal feed as an anti-caking agent.

When silica is exposed to high temperatures, either through human-directed processing or from volcanic activity, it becomes crystalline. Crystalline diatomaceous earth is used in manufacturing processes, such as filtration. Unlike its amorphous cousin, crystalline diatomaceous earth has been shown to cause lung damage with long-term exposure. It also has no effect on the exoskeletons of insects or parasites and thus has no beneficial use in the chicken yard.

Should you decide to use diatomaceous earth as part of your parasite prevention program, make sure to always opt for the food-grade kind. Food-grade diatomaceous earth must contain less than 1 percent crystalline silica to be labeled as such. Under no circumstances should you go for the big, cheap bag of diatomaceous earth intended for use in pools. It may seem like a good deal, but it is crystalline and not only is it dangerous for you and your chickens but it will also do nothing to prevent parasites.

In my opinion, it's truly unfortunate that diatomaceous earth has gotten such a bad rap. The original alarm was probably sounded due to the possible trace amounts of crystalline in the food-grade kind. But think of it this way: It's *long-term* exposure to crystalline diatomaceous earth that has been shown to cause lung damage. In the course of your chickens' lifetime, a tiny amount of possible exposure is unlikely to cause an issue.

Still, it is true that there is a minuscule risk, and you will need to decide if it is worth taking or not. When weighed with the risk of mite and lice infestation or the use of other chemical parasite treatments, I'll take good ol' diatomaceous earth.

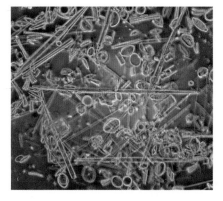

Diatomaceous earth is an insecticide and therefore can be harmful to bees and other beneficial critters in your backyard. Limit its use to dust baths and the coop and on the chickens themselves if direct treatment is necessary.

# SUPPORTIVE CARE

## for When Disease Strikes and Emergencies Happen

Chickens are susceptible to a wide variety of diseases, but you and your flock will never come in contact with the vast majority of them. Some chicken illnesses are serious, some can be managed with veterinary care, and many can be overcome by supporting your ailing bird at home with a little old-fashioned TLC. In this chapter, I list general tips for supportive care that are easy to remember. You may already have some of the supplies I mention around the house, but if not, they are easy to find.

## No Matter What Happens, You Are a Good Chicken Keeper

Despite your best efforts, at some point in your chicken-keeping journey you will more than likely be confronted with at least a mild case of illness in your flock. I receive numerous messages and emails from people who take top-notch care of their birds, and yet they are facing illness and/or parasite infestation. Even though their chickens have a better life than most chickens in the world, the main reaction these distraught folks convey to me is guilt. They feel like it's their fault that their chicken is sick or has parasites. "I don't know what I did wrong," they write. It makes me sad because they are caring and compassionate people and their chickens are very fortunate to have them.

I want you to know that most of the time when a chicken becomes ill, if you have your best practices in place, then it likely has nothing to do with anything you did or didn't do. The unfortunate fact is that baby chicks sometimes just aren't meant to survive (that's why Mama Hen produces such a large *clutch*—the number of eggs she intends to hatch), and considering their often stressful first days of life, it's a miracle how well they usually do. When an adult chicken dies, especially if it's a laying hen, it may help you to remember that in general, modern chickens are bred for production rather than longevity. So, no matter what you and your chickens

---

### Supplies to Have on Hand

You can purchase the electro-lyte, vitamin, and probiotic (EVP) powder as well as the Nutri-Drench (a concentrated vitamin supplement made for chickens) and other concentrated chicken vitamins online or at your local farm store. Bach Rescue Remedy is available at most pet supply stores, and you can purchase homeopathics at most health food stores or online.

+ electrolyte, vitamin, and probiotic powder for chickens (or homemade electrolytes, page 121)

+ green tea bags

+ needleless syringe

+ Bach Rescue Remedy or other similar flower essence combination

+ paper towels

+ blow dryer with low setting

+ Nutri-Drench or other concentrated chicken vitamin

+ homeopathics in 30c potency (see list on page 147)

are faced with, rest assured you are doing a great job! They are so lucky to be cared for by you, even if it's just for a little while.

## The Sick Chick Protocol

As you may remember, baby chicks have about 48 hours to eat and drink after they hatch. If they fail to do so within this time frame, they are in danger of developing a condition known as starve-out. Starve-out is basically when a youngling is so undernourished that it becomes weak and loses the will to survive. The chick will stand in one spot, either staring into space or down at the ground. Its wings will be droopy. It will not seek out food or water. As you can imagine, this is a very sad state of affairs both for the baby chick and for whoever is looking after it.

Starve-out affects chicks that have been shipped through the mail at a much higher rate than those raised by a mother hen. The reason for this is obvious when you consider everything a shipped day-old chick goes through right after hatch. That amount of stress is a challenge for such a young one to overcome, especially if they are a tad weaker than their flockmates. Starve-out is very common, and sometimes there is simply nothing that can be done. However, you can try feeding the chick a homemade slurry at intervals throughout the day in an effort to save it. The formula on page 134 is helpful in many conditions where a baby chick is struggling to survive, even if you're not sure what's wrong.

After a few days of being fed the slurry formula, your baby chick will hopefully gain enough strength and resilience to begin to eat and drink on its own. I've witnessed baby chicks that seemed to be on the brink of death brought back using this technique. I've also witnessed times when nothing worked, and the chick did not survive. Since vulnerable chicks will sometimes get picked on by other members of their flock, it's best to separate them while they're being treated. If possible, give the chick one or two gentle friends for company and watch them to make sure the sick chick is not being picked on. Chickens are flock animals from the start, and baby chicks will get very lonely and stressed out if left completely alone.

# How to Make the Sick Chick Slurry

You can make the following formula when you have a sick chick who is struggling, then hand-feed the slurry to your baby chick via a needleless syringe several times a day.

## WHAT YOU NEED

* 1 green tea bag
* 1 cup tepid (*not* hot) water
* EVP powder (or homemade electrolytes, page 121)
* ¼ teaspoon raw honey
* 1 egg yolk
* small needleless syringe
* 2 drops Bach Rescue Remedy or other similar homeopathic flower essence (optional)

## DIRECTIONS

1. Steep the green tea bag in the water for about 1 minute to prepare a weak tea.

2. Mix in the appropriate amount of EVP powder per the package instructions. Alternatively, add a pinch of your homemade electrolytes and the raw honey.

3. Add 2 tablespoons of the tea mixture to the egg yolk and stir. (You can save the remaining tea mixture in the fridge and use within 24 hours. Always bring to room temperature before administering.) Add 1 drop of Rescue Remedy, if desired, or place 1 drop of Rescue Remedy on the chick's back each time you feed it the slurry.

4. Place drops of the formula on the side of the chick's beak so that the chick can lap it up. If they are not drinking it once it pools inside their beak, they are likely too far gone to be saved without the help of a licensed veterinarian, and even then it's unlikely they will make it. Do not force-feed, as doing so can easily cause a chick or chicken to aspirate food or liquid. Repeat several times a day.

## The R.E.S.T. Method

If you find an adult chicken listless, staring into space, hunched over, or not eating, he or she probably isn't feeling very well. Since chickens are such experts at hiding illness because they don't want their flock mates noticing that they are weak, you may not realize they are sick until their illness has progressed to the point where they can't hide it anymore, in which case it's in your best interest to act fast. Since I'm not the calmest person under pressure, I created the R.E.S.T. method to help myself and others through these situations. It's easy to remember, and it can be a powerful tool for returning your chicken to a healthy state. After you've given your chicken a once-over to check for any obvious problems or injuries, follow the R.E.S.T. method:

> **R**—Remove from flock
>
> **E**—Electrolytes, vitamins, and probiotics
>
> **S**—Scrambled eggs
>
> **T**—Temperature control

You can use this method for most illnesses, or when you have no idea what's actually going on with your chicken. You can also use this method when your chicken has had a stressful experience or when they

---

The R.E.S.T. method can be used in times of illness, injury, and even shock.

have been injured. Of course, the best scenario would be for you to seek treatment from a licensed veterinarian, but that's not an option for many people. Here's a breakdown of what you can do and why I recommend each step.

**Remove your chicken from the flock.** Having one sick chicken doesn't automatically mean your whole flock will become infected, but it's still a wise practice to isolate birds that are showing signs of illness. This is not just to stop the spread of disease but also to protect your weaker flock member from the unfortunate nature of the pecking order, which makes sick chickens vulnerable to attack. Additionally, just like humans, chickens need rest to recover from illness. Being in a nice quiet place such as your garage, a bathroom, or other temperature-controlled area can help them do that.

**Electrolytes, vitamins, and probiotics.** When I got sick as a kid, my mom used to give me Gatorade (and chocolate pudding, but that's another story). She did this because she knew that when children don't feel well, they often refuse to drink the liquids they sorely need. Chickens are also known to shun food and water when they feel lousy, which makes them vulnerable to weakness and dehydration. Giving sick chickens EVP water or homemade electrolyte water (page 121) will encourage them to drink and hopefully give them the boost they need to regain some energy and appetite.

**Scrambled eggs.** While there are occasions when a chicken's digestive system needs a break so that their body can focus on getting well, chickens that don't eat for a long time can literally lose their will to live. There are few foods as tempting to chickens as scrambled eggs, and it will give them a good dose of protein, fat, and other nutrients to aid in their recovery.

**Temperature control.** When chickens aren't feeling well, their bodies need to be working on healing, not trying to stay warm or cool. For this reason, I suggest bringing them inside your home or another enclosed building like a garage, where you can regulate the temperature. If you must use supplemental heat, make sure your chicken can get away from it so that they don't get overheated.

While the R.E.S.T. method is unfortunately not a magic pill, it will hopefully have a positive effect. Sometimes, all a chicken needs to recover is some TLC and

---

### Bach Rescue Remedy

Note that you can also rub two drops of Bach Rescue Remedy on your chicken's back or put two drops in their water when they are sick, injured, or stressed.

a comfortable place to rest. At the very least, it can buy you some time while you evaluate the issue and make a plan for veterinary care if necessary and possible.

## The Most Common Chicken Ailments

After many years of helping people with their chickens, I've found that there are certain ailments that occur more than others. While there are many different chicken illnesses and conditions, the following are what I feel are the most common. If you're looking for a more in-depth guide, I recommend reading *The Chicken Health Handbook* by Gail Damerow.

### SNEEZES, WHEEZES, AND GURGLES

Several different respiratory ailments can affect chickens; some are very serious, and others are manageable and relatively easy to treat. Unfortunately, when you hear that first sneeze, wheeze, or gurgle, you won't know how serious it is until you get a proper diagnosis from a veterinarian, and sometimes even vets are simply making their best guess.

I've rescued many chickens over the years and have consequently dealt with

Poor coop ventilation elevates the risk of chickens developing respiratory disease during winter months.

my fair share of sniffles within my flock. I'm going to share the supportive care that I have found works best for me, but if you find that these actions aren't helping, and certainly if you have more than one chicken die or become seriously ill, veterinary care is in order. If a vet just isn't possible, you may consider having a necropsy (postmortem examination) done through your state health department so that you can understand why the chicken died and may be better able to deal with

## Chicken Respiratory Illness Protocol

✤ Administer the R.E.S.T. method to any sick chickens at the first sign of illness.

✤ Add one cut-up clove of garlic per gallon to the sick chicken's water as well as to the flock's water, or mince a clove and add it to their feed or scrambled egg.

✤ On two paper towels, place a few drops of one or more of the following gentle essential oils: eucalyptus, lavender, tea tree, peppermint, lemon, and thyme. Hang one paper towel in the coop with your flock and one near your sick chicken.

✤ Add a generous sprinkling of dried oregano and thyme to scrambled eggs or to your flock's feed. You can also add the herbs to the sick chicken's feed or to scrambled eggs, but discontinue if the change in their food stops them from eating.

✤ Add apple cider vinegar (ACV) to your flock's water to help prevent spread and for its immune-boosting properties. (ACV can be used with garlic.)

✤ Optional: Add a couple of drops of Bach Rescue Remedy or similar flower essence daily to the sick chicken's water and to the rest of your flock's water. (Flower essences can be combined with ACV, but it may not work as well if added to garlic water; add garlic to feed if using flower essences in water.)

✤ Consider administering the homeopathic remedy aconite in 30c potency to the sick chicken at the first sign of illness. Administer other appropriate homeopathics as needed. (See page 147.)

the problem in the future. Necropsies are usually much cheaper than a veterinary visit, and the information they offer can be valuable to the future health of your flock.

I'm often asked why I don't recommend dosing a whole flock with antibiotics when respiratory illness hits, unless directed to do so by a licensed veterinarian. Unfortunately, the frequent overuse of these lifesaving drugs has led to widespread antibiotic resistance. As a result, most of them have been pulled from the shelves in the United States, where they were once available over the counter. As we've discussed, chickens aren't the only ones hurt by medication overuse; it hurts every living thing in the ecosystem we all share. If you prefer to give your flock a natural alternative to antibiotics, there are a few concentrated oregano products on the market made specifically for chickens that claim to help with respiratory conditions. Alternatively, you can dose your flock with 1 tablespoon of colloidal silver per gallon of water for a few days until the risk of illness has passed.

## Are Essential Oils Safe?

I know essential oils are controversial, possibly due to zealous folks claiming they cure everything from a stubbed toe to cancer, but I find them helpful for certain situations and recommend using a few gentle ones, such as eucalyptus, lavender, tea tree, peppermint, lemon, and thyme, when the threat of respiratory illness hits a flock. I also use them during the coldest parts of the winter, when my chickens are at elevated risk of respiratory problems, to give them an extra boost and keep those sinus passages open.

I never put undiluted essential oils directly on my chickens or administer them internally. That said, you can buy premade essential oil products (similar to a "vapor rub" for human use) from an online vendor or your local farm store. These products are fairly diluted, and you can safely put them directly on your chickens if you follow the directions.

## PASTY BUTT

Baby chicks that have been stressed, have become too cold or too hot, are sick, or have not received proper nutrition can develop a collection of fecal matter around or over their vent. This condition is called pasty butt, and it is very common among shipped chicks. If left untreated, pasty butt can make it difficult or impossible for a chick to expel fecal matter, which, I'm sure you can imagine, has terrible consequences. Fortunately, the remedy for pasty butt is very simple. Here's what you need to do.

1. Clean off the poop with a warm, wet paper towel. If the case is severe, you can run their bum under warm water. *Never pull off the fecal matter. Ease it off gently by wetting it down and breaking it apart.*

2. Thoroughly dry off your baby chick with a towel or blow-dry them on the lowest setting.

3. Place the chick back under the heat.

4. Check any affected chicks daily, as pasty butt has the tendency to recur.

If you like, place a drop or two of Bach Rescue Remedy on their back before treatment to make the experience less stressful for them. And of course, make sure your baby chicks are getting the proper nutrition. Starting them out with either ACV or EVP, as well as fermented feed, can also be helpful.

STEP 1

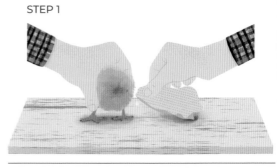

STEP 2

STEP 3

## COCCIDIOSIS

Despite your best practices and efforts to gradually expose your chicks to their outside environment, and even if they are on medicated feed, it is still possible for your flock to come down with an active case of coccidiosis. Here are some signs to look out for.

✤ You've recently brought new members into your flock, and you suddenly have a sick chicken.

✤ You've adopted older baby chicks and one or more appear slumped over and unwell.

✤ You have a chick or chicken that looks unwell, and you've recently discovered blood in their feces. (Note that some occasionally red-tinged stool is normal and is the result of the intestinal lining being shed.)

If any of the above situations occur in your flock, your best course of action would be to have your local veterinarian's office perform a fecal test. Fecal tests alone do not always guarantee an accurate diagnosis, but they provide information that can be helpful to you and your veterinarian when determining a treatment plan.

If you can't get a fecal test, my advice is to treat your whole flock with amprolium. Follow the directions carefully, and do not administer it for longer than is required. When your birds are done with their course of amprolium, put them on a 2-week course of a high-quality chicken vitamin that has thiamine as a top ingredient. If you do not have access to chicken vitamins, add 1 tablespoon of nutritional or brewer's yeast per gallon of their feed daily for 2 weeks. (If you ferment your chicken feed, sprinkle on the yeast after it is fermented and right before feeding.) It's important that you don't skip this step, as it may help you avoid future issues caused by thiamine deficiency.

## OTHER INTERNAL PARASITES

If you observe worms in your chicken's poop (this is super gross but make sure they aren't maggots, which look like small worms but are actually fly larvae) or if you have one or more sick chickens and you believe an infestation could be the culprit, the first rule is this: Don't panic. Take a deep breath and remind yourself that this is part of nature and that you are being given an opportunity to reevaluate your practices and improve on them as needed. The best option would be to have your local veterinary office test a fecal sample to determine what parasite you are dealing with. Once you have this information, it will be easier for you and/or your veterinarian to determine what route to take, as different parasites require different medications.

If you want to go the natural route, there are a few herbal dewormers to choose from in the United States. The strongest ones will contain wormwood, but be sure that the company doesn't recommend it be used daily. Even though wormwood is an herb, it is powerful and can have adverse effects if given long term with no breaks in treatment. The best companies will recommend a short course of wormwood along with other soothing and cleansing digestive herbs.

While I'm a big believer in the efficacy of homeopathics, I don't recommend using a homeopathic remedy as your sole treatment for internal parasites. Instead, use one in conjunction with an herbal formula or prescribed medication for added support. I recommend the homeopathic cina in 30c potency. Simply add two pellets (or two drops if it is in liquid form) to their water for 3 days at the beginning of their treatment.

If you live on a small lot and your chickens are in a stationary coop and run, you may need to consider adding an herbal parasite prevention program to your chicken-care regimen. It's possible that my garlic, oregano, and thyme recommendations (page 122) will do the trick, but it's also possible that the parasite load in your soil will become so high over time that your flock will need something stronger. Most of the herbal dewormers on the market offer periodic dosage programs that are easy to administer. If you just can't seem to get ahead of the issue, it may be time to have a fecal sample done and administer conventional medication. After that, you can confidently return to natural preventives, and with a clean slate.

## THE EGG-BOUND HEN

When pullets reach laying age, their reproductive system plays a critical role in their lives. Because of this, it is essential that layers get the proper nutrition they need for their reproductive cycle to remain in working order. However, since laying hens are usually bred for high production, even with the best care they can sometimes still end up with a laying issue. One of the most common egg problems is egg binding, which means that an egg is stuck in a hen's oviduct and she is having difficulty passing it.

An egg-bound hen may walk with her legs farther apart than normal and/or her backside close to the ground, kind of like a penguin. She may be listless, standing in the corner, and have a pale comb and wattles. She may also appear to strain her vent like she needs to pass something. If you find your chicken obviously ill, one of the first things you should do is check to see if you feel an egg near her vent. Be very gentle. The last thing you would want is a broken egg stuck inside her, which can cause a fatal infection. If you feel a bulge near her vent, then she is likely egg-bound.

The longer it takes for a chicken to pass an egg, the more danger she is in, so act quickly to try to encourage laying. Start with the R.E.S.T. method: Get her inside, where it is warm and comfortable. While she is enjoying some scrambled eggs (for egg-bound hens, crush and scramble the clean eggshell as well), run a warm bath for her in a sink, bathtub, or plastic tub, and add some Epsom salts. You can also add a single drop of lavender essential oil to the bath for ultimate relaxation.

Place your hen in the warm Epsom salt bath for up to 20 minutes. Hopefully, the warm water will relax her vent enough to pass the egg. If not, dry her well, rub a little bit of coconut oil or Vaseline over her vent, and return her to a warm, quiet spot. It's very important for her to be in a warm area, but if you have her under supplemental heat, make sure she can get away from it to cool off if needed. You can bathe her two or three times a day, as long as she isn't getting stressed from the activity, which would actually discourage laying rather than encourage it.

If your chicken still hasn't passed the egg after 24 hours, try offering her a calcium supplement for birds, usually marketed for parrots. Be mindful to follow the directions carefully concerning dosage. If that doesn't work and she still hasn't laid in more than 48 hours, she really needs veterinary care if that is an option. You could attempt to remove the egg with a gloved and lubricated hand as a last resort, but it could possibly kill your chicken or introduce infection. Whatever happens, remember that your chicken had a wonderful quality of life, and that's what matters most.

A warm Epsom salt bath can sometimes relax a hen's body enough to release a bound egg.

## MITES AND LICE

It's possible to have the best care practices and still end up with mites or lice in your flock. If you are faced with an infestation, you can choose from several options that will have relatively little effect on the environment compared to more conventional treatments. Still, it's important to mindfully apply the remedy because any insecticide—whether it is natural or not—has the potential to harm wildlife.

To apply mite and lice treatment, concentrate on the back of the neck, under the wings, and around the vent.

Because mite and lice infestations happen to vulnerable birds, make sure to check each member of your flock for injury or illness as you treat them. You must also thoroughly clean and treat your coop each time you treat your chickens so that you can properly halt the life cycle of the parasite. Make absolutely sure the pesticide gets into all the nooks and crannies, as well as onto roosts and inside nesting boxes. Always follow product directions carefully.

**Spinosad** is a substance produced by bacteria that live in the soil that is toxic to some insects, including mites and lice. It comes in a concentrated liquid form made specifically for chickens that must be diluted and then sprayed onto each bird, as well as your cleaned coop. It can be harmful to bees when wet but is said to have no effect on them once dried. At this time, spinosad is unfortunately very expensive, though it does last a long time and usually requires only one application. If you feel you need to administer it again, do so in 10 days.

**Pyrethrum and permethrin (pyrethroids)** are insecticides made from the active ingredient in chrysanthemum flowers, called pyrethrin. Permethrin differs from pyrethrum in that it has been modified to last longer and be more effective. Permethrin is technically not a completely natural product, but it is fairly benign and so is used by many people who wish to cause as little damage to the environment as possible. Still, pyrethroids are toxic to bees, aquatic life, and cats. Some research even suggests they may affect human and animal endocrine systems. Take care to limit the use of pyrethroids to the chickens and their coop, and reduce your exposure by using gloves and a mask that prevents product inhalation.

Pyrethrum and permethrin are available in sprays and powders. Be sure to use only the kind that is labeled for poultry and follow the product directions carefully. Apply to each bird and your clean coop twice, 10 days apart. You may need a third application, depending on the level of infestation.

**Food-grade diatomaceous earth**, in my experience, also works to treat infestations of mites and lice, though many people say it does not work for them. You do need to use it correctly, and it usually requires more treatments compared to other remedies to be effective. It's best to administer three treatments, 10 days apart. You will also want to clean your coop every time you treat the chickens and dust every nook and cranny, the nesting boxes, and the floor as well. While food-grade diatomaceous earth carries very little risk for lung damage with such limited exposure, it's best to wear a mask when working with any dust-like material. Diatomaceous earth can also harm bees and other beneficial insects, so be mindful of using it only where needed.

## Using Homeopathy on Your Chickens

The words *holistic* and *homeopathic* are often used interchangeably, but homeopathy is its own all-natural modality that was discovered in the late 1700s by a German physician named Samuel Hahnemann. The basic principles of homeopathy are that "like cures like" and that the body is more than capable of healing itself. Each homeopathic remedy contains a minuscule amount of a substance that in larger amounts would induce the same symptoms it is looking to address. Ingesting a small amount of the substance—so small it is considered to be at an energetic level— is thought to prompt the body to work against whatever is ailing it. Homeopathics can be in liquid, pill, or pellet form, and come in different potencies. The potency I usually recommend for chickens is 30c, which is best suited for acute situations.

I was first introduced to homeopathy when I worked at the holistic animal supply store in Arizona. I've used it under the direction of naturopaths and various alternative practitioners and now do so with the guidance of my family's licensed homeopath. I have witnessed time and again the right homeopathic remedy work wonders not only on my chickens and ducks but also on my husband, my children, my dogs, and me. Because of my positive experience, I am sharing with you a list of remedies I feel are good to have on hand in your chicken first-aid kit, along with their common indications. These remedies are most often found in health food stores or online, in pellet form. Each homeopathic remedy is used for many different ailments, so don't worry if you see indications other than what you are using it for listed on the label.

Homeopathy is often criticized as ineffective, but this usually has more to do with how it's administered (selecting the wrong remedy or dosing incorrectly) than with the modality itself. Unlike conventional medications, which react to specific conditions and/or symptoms, homeopathics react to the nature of the individual and how they, specifically, are experiencing an illness or injury. Homeopathy is a complex modality that requires years of practice or schooling to truly master. That said, the remedies I recommend are widely used for the common ailments I mentioned. For complicated situations or to find the most appropriate remedy match, it's best to seek the advice of a licensed homeopath.

### HOW TO DOSE HOMEOPATHICS

Because homeopathics work on an energetic level, adding one or two pellets or drops to liquid will turn any amount of that liquid into one dose. For instance, if you add two pellets of arnica in 30c potency to a glass of water, anytime you sip that water you are receiving a dose of 30c arnica.

To give your chicken a proper homeopathic dose, dissolve one or two pellets (or two drops if in liquid form) in a small glass of room temperature water. Siphon some of the water into a needleless syringe. To give one dose, place a few drops on the side of your chick or chicken's beak and let it drip into their mouth. Because of the energetic nature of homeopathics, even if they do not actively swallow, any amount that enters the mouth is considered a proper dose. Administer two or three doses, depending on instructions, 10 minutes apart, and watch carefully for changes.

If your chicken perks up at any time during or after dosing, stop administering the homeopathic and don't dose again unless they backslide. If you don't observe any improvement, you can dose again in a few hours and observe. If your chicken shows no sign of recovery (even the tiniest bit of perking up is considered a positive sign) or seems to be getting worse, you are probably not using the right remedy.

## When Emergencies Happen

This would be a very long book if we went over every single scenario of chicken injury, so instead, I will give you a list of emergency items you should have on hand, as well as a quick overview of what actions to take when a chicken is injured or in shock. If you want a more in-depth look at chicken health, you can check out Gail Damerow's book *The Chicken Health Handbook*.

If one of your chickens has become injured or wounded, first separate them from the flock. Since injured birds are usually pretty stressed out, I recommend rubbing two drops of Bach Rescue Remedy or a similar flower essence formula on their back immediately and repeating this action a few times a day until they recover. You can also add two drops of flower essence into their water so they can get a calming dose each time they drink.

---

## Homeopathic Remedies Helpful for Chickens

**Aconitum napellus (30c):** Use for sudden fear or shock, a sudden chill, heat exhaustion, trauma, or illness that comes on suddenly. It is often the first remedy that is used when a chicken is sick. Give two or three doses right at the onset of the issue, 10 minutes apart.

**Antimonium crudum (30c):** Use for a chicken with obvious respiratory issues, specifically if there is rattling in the nasal passages or chest. Give three doses, 10 minutes apart.

**Arnica montana (30c):** Use for shock, injury, bruising, swelling, and pain. Add two pellets to their water, and allow them to drink freely until you see progress. If they are not drinking voluntarily, give two or three doses 10 minutes apart.

**Arsenicum album (30c):** Use for ease in dying, especially if the chicken is in distress, panicking, or seizing. You can also give it when there is accidental poisoning or food poisoning such as botulism.

Give one or two doses, 10 minutes apart, every few hours as needed.

**Carbo vegetabilis (30c):** This is often used as a last resort if your chicken appears to be extremely ill, has curled toes, and is near death. Give one dose and wait to see how they respond. It can also be used for some respiratory issues, in which case, give two or three doses, 10 minutes apart.

**Gelsemium sempervirens (30c):** Use as a supportive treatment for a chicken who seems to have "given up," is lethargic, won't eat, and is depressed. You can add it to the Sick Chick Slurry (page 134). Give two or three doses, 10 minutes apart. You can also give it to chickens who have coccidiosis or egg-bound symptoms, namely when they are hunched over and listless.

**Cina (30c):** Give during treatment for internal parasites. Add two pellets or drops in the water for 3 days at the beginning of treatment.

**A chicken that is in shock** will stand still, stare into space, and refuse to eat or drink. They really need to start eating and drinking within 24 hours or they will be in grave danger, and you might lose them. If your chicken goes over a day without food and water, consider hand-feeding some raw egg yolk mixed with electrolyte, vitamin, and probiotic water (or homemade electrolyte water, see page 121) via a needleless syringe.

If a chicken has had a terrible fright or is in shock, I will immediately administer the homeopathic aconite in 30c potency via needleless syringe. You can also add two pellets or two drops of homeopathic arnica to their water, which will also help with pain and bruising. See page 147 for more detailed information on homeopathics.

To avoid aspiration when medicating via syringe, place a few drops of liquid on the side of your chicken's beak so that it can drip into the mouth and activate the drinking reflex.

## Flower Essences

Flower essences are a nontoxic, natural modality that was discovered in England in the early twentieth century by Dr. Edward Bach. Dr. Bach believed that the energy of different types of plants and flowers can help with the emotional challenges of humans, animals, and even other plants. Whenever I suggest their use, I get pushback from those who feel there isn't enough evidence to support their efficacy. I understand this criticism, but I have seen them work and so my recommendation stays.

Flower essences can be found online and in pet stores and health food stores. Flower essences are similar to homeopathics, in that they are believed to also work on an energetic level. They come in pellet or liquid form.

**If your chicken is wounded,** first make sure your hands are clean, then try to stop any bleeding by dabbing the wound with a clean, damp cloth. If bleeding won't stop, use styptic powder, and wait a few minutes for it to dry. Next, you will need to fully clean the wound. You can do this with gentle soap and water and a clean cloth or sponge, or saline in a needleless syringe. If the wound is superficial, spray colloidal silver or an antimicrobial on the clean wound and allow it to air-dry. If the cut is deeper, apply a thin layer of antimicrobial ointment and consider covering it with gauze, but only if you can keep a good eye on the chicken to make sure they don't peck at the bandage and get the wound dirty.

**If your chicken has an obvious broken leg,** contact a licensed veterinarian if possible. If you can't access or afford a vet, don't cast out all hope. You might be able to splint the leg using a wooden craft stick, a strip of stiff cardboard, or a small stick. Chickens are surprisingly resilient. If you can get them past the initial shock of injury, they can often maintain a wonderful quality of life with compassionate care, even if they have a permanent scar or limp.

---

## Basic Chicken First-Aid Supplies for Injuries

Note: While I generally favor natural remedies, I also recognize that modern medicine is great at handling wounds and injuries. For this reason, not everything on the following list is completely natural.

**Must-Haves:**

✤ antimicrobial ointment (without pain relief) or raw honey

✤ clear bandage tape

✤ colloidal silver spray or antimicrobial spray for chickens

✤ gauze pads

✤ gauze wrap

✤ saline

✤ scissors

✤ sterile gloves

✤ sterile syringe

✤ styptic powder or cornstarch

✤ tweezers

✤ vet wrap

✤ wooden craft sticks

**Optional:**

✤ Bach Rescue Remedy

✤ homeopathic aconite 30c

✤ homeopathic arnica 30c

# Humankind's Most Amazing Common Denominator

Oh, to be a kid again and believe all dreams are possible. As a child, I sincerely thought that someday I would help to create a better, more unified world. On my imagined planet there would be no war, no injustice, and no nuclear bombs. No one would starve. Everyone would have water. I think most kids believe these goals are attainable because to them, the alternative is unimaginable. But then we grow up, and as the cruel reality of the human condition sinks in, we may find ourselves

laughing at the absurdity of such childish ideals.

This transition from idealist to so-called realist was not easy for me. As a young adult, I traveled all over the country in search of . . . something. At the pinnacle of my odyssey, when I was living in Los Angeles and fruitlessly looking for work as a writer for television and movies, I took a trip with my dad to Guatemala. The civil war was over, and he had a serious surgery coming up. He wanted to visit his homeland because he didn't know if he would be able to travel there again. I teased him about how melodramatic he was being, but deep down I was scared that this would be the only opportunity I had to travel to Guatemala with my dad. It turned out I was right.

Despite the ravages of war, Guatemala has areas that are so strikingly beautiful they border on magical. At one point in our trip, my dad and I, plus two of my uncles and my grandmother, drove to the coast to visit some family property. The view from my uncle's compact car was breathtaking: there were volcanoes on the horizon, Mayan villages, and jungle as far as the eye could see. My grandmother told me stories of my dad's considerably rebellious teenage years while my uncles laughed and translated. My dad sat next to me in the backseat, looking sheepish, resigned, and perhaps a bit proud. "Aye, Mama!" he said over and over.

It was interesting to see the dynamic between my dad and my grandmother. He was such a strong figure in my life, and yet when he was around her, I caught glimpses of a young boy who still needed his mother's approval. And the apple, or perhaps the avocado, didn't fall far from the tree. While I cherished every moment of this trip with my dad, there were contentious times due to our old, stubborn patterns. As we were getting out of the car at our first stop, we quarreled over something trivial that we both blew out of proportion. As a result, we didn't talk to each other much while we strolled through the streets and took photos. Once back on the bumpy road, we remained almost silent until we finally arrived at the coast.

The heat was wet and heavy, and jungle surrounded the small plot of land. After a sweaty tour of my uncle's house, I wandered alone into a nearby garden to steal some fresh air and a few moments from trying to follow conversations with my limited Spanish. The garden was lush and overgrown with flowers and vegetable plants. Huge butterflies danced from stem to stem.

As I strolled through the grounds, I was surprised to come upon a mother hen with baby chicks in tow. Having spent my entire childhood in the suburbs and then living in cities as an adult, such discoveries were a novelty. While I watched them with adoration, one of the chicks fell and struggled to get up. It called out in distress to its mother, who hadn't yet realized the situation. I don't think I had ever handled a chick before, but I swooped down, picked it up, and set it gently upright. It scurried back to its mother, who took it under her wing.

My dad wandered from the house. I called to him, "Daddy, look! Baby chicks!" The two of us stood for several minutes beside each other, watching the mother and babies in silent delight. Our cultural and generational barriers slipped away.

# AS YOU GROW, SO DOES YOUR FLOCK

As you become more confident in your chicken-keeping abilities, you will likely begin to consider adding to your flock. After all, what could be more irresistible than new baby chicks and more eggs to feast upon, sell, or share? Expanding your flock can be a wonderful experience for you and your family, and it could even benefit your flock's dynamics for the better. That said, adding new chickens to an existing flock takes patience and time, and there are a few things you will want to consider before taking the plunge.

## Quarantine *If You Can*

Just as you took care to build your chickens' immunity to the bacteria, fungi, viruses, and parasites in their environment, you will need to take the same care to build the immunity of your new arrivals. You will also need to consider that new chickens could bring in pathogens that are unfamiliar to your established flock.

It's generally understood that chickens coming from bad situations are more likely to have health problems. But even chickens that are well cared for and loved can have hidden conditions that aren't visible to the naked eye. A hen can appear perfectly healthy and yet have an infestation of mites, lice, or internal parasites. She could be carrying diseases your flock isn't immune to, such as a foreign strain of coccidiosis, or a chronic respiratory disease that goes unnoticed unless she is under stress. Knowing this, it's just plain common sense to take precautionary measures before exposing your flock to new members. This includes a short quarantine and other proactive steps for both new adoptees and your existing flock.

If you have the space and means, quarantine any new chickens brought into an existing flock for at least 2 weeks. If you are in an urban or suburban environment and aren't able to quarantine properly, don't worry and just take the other precautionary measures I list in this chapter. Despite the risk, I truly believe the benefits of adding to your flock are worth it both for your family and your birds.

To quarantine, find a spot as far away as possible from your chicken yard where you can place a separate, secure enclosure safe from predators. You want there to be enough distance from your existing coop

---

### The Reality of the Pecking Order

If you already have chickens and have been watching their behavior, you've likely witnessed the peculiar nature of the pecking order. Even tiny chicks will sometimes play in a manner that mirrors adult posturing and fighting. New chicken keepers might find this concerning because they think it indicates they have a rooster or roosters in their brood, but rest assured that both male and female youngsters can behave in this way, and it just means they are practicing for future endeavors.

Full-on pecking order behavior doesn't usually begin until about 3 to 6 weeks of age, and most often at the later end of this range. Once the ranks have been determined, however, any new chickens that are added to the flock can be injured,

so that you can practice good biosecurity (actions taken for the purpose of disease prevention) between the two areas. This is generally recommended to be at least 33 feet (about 10 meters), but if you don't have enough room to meet that requirement, just do the best you can. Make sure to wash your hands before and after you have contact with either flock. Wear different shoes each time, cover your shoes with plastic film, or clean your shoes between flock visits. Make sure the two flocks have separate everything: separate space, separate feeding area, and separate

and as chickens reach adult age, the pecking order can even mean life or death.

In an established adult flock, there will be chickens at the top of the pecking order and chickens at the bottom. Your flock will develop its own personality as a unit, according to who is in charge, who will be allowed into the upper circle, and who will be excluded from certain privileges. This aspect of chicken nature can be difficult for sensitive chicken parents to witness. Why is Bertha keeping June from the food bowl? Why does Rosa always have to eat last? How come Tisha seems to always peck and scratch alone?

It's normal and healthy for humans to feel empathy when someone is left out. In nature, we're much more like our benevolent cousins the bonobos than we are chickens. Because of this, we sometimes feel an urge to correct chicken behavior in an attempt to have it mirror what we consider acceptable. I completely understand this, but I encourage you to try observing the world through the eyes of your flock.

For chickens, the pecking order is natural. It's their instinct to take their proper place because they experience life as a flock, not as an individual. The hierarchy, though it seems harsh to us, exists to ensure their survival as a group and a species. As long as no one is getting hurt or being kept from food and water for extended periods, I generally leave my chickens to their natural tendencies. There are times, however, when I need to step in to make sure no chicken gets hurt. One of those times is when new additions are introduced to a flock.

watering area. This will give you the best chance of avoiding cross-contamination.

After the 2-week quarantine period is over, you will need to begin the process of combining your two flocks. It's important to note that quarantining is not foolproof. Certain pathogens can be sneaky, and chickens can hide illness very well, even under stress. Taking in any new bird comes with a level of risk. I've decided that risk is worth it, and so have many other chicken keepers around the world.

## MITIGATING STRESS

Whether you're a chicken or a human, change is tough. Imagine living happily in your home with your spouse and kids, then suddenly having your in-laws move in with you. Or imagine if you were pulled from a happy home, then transported in

some sort of alien vehicle to a commune you're expected to live in for the rest of your life! Even chickens that are well cared for can find it difficult to change homes or have strangers move in. Add in the possibility of trauma, and big changes could push a chicken into illness, parasite infestation, or behavioral problems.

There are some steps you can take during the quarantine period (or if you're not able to quarantine, during your new chickens' first 2 weeks with their new flock) to give them the best chance of a smooth transition.

✤ Begin the immune-boosting practices listed in Chapter 8 with your newbies if you already have them in place with your established flock.

✤ Treat your new members for external parasites right away, even if you don't observe any critters. This is extremely important.

✤ Cook up some comfort food like scrambled eggs and add some minced garlic, oregano, thyme, or other gentle medicinal herbs for a tasty treat and immune booster.

Sorting out pecking order issues can be stressful to chickens.

Consider getting a fecal test done with a veterinarian if internal parasites are suspected, or administer an herbal dewormer that contains wormwood (just remember to follow the directions carefully and avoid overuse).

If you discover that a newly adopted chicken is sick, go ahead and follow the R.E.S.T. method (see page 135) or the Chicken Respiratory Illness protocol (see page 138) and/or seek veterinary care as needed. Unfortunately, you will also need to do some soul searching, because you have a genuine dilemma on your hands. The harsh reality is that even if the chicken is able to fight off whatever is ailing them, there are diseases that chickens can carry and possibly spread for a lifetime, even if they no longer suffer from them. Since my flock is mostly rescues, I've made an agreement with myself and my chickens that I'm willing to absorb the risks associated with bringing newbies home, even sick ones. Keep in mind, however, that I am a seasoned chicken keeper who is versed in risky situations. I also have access to a great avian vet. In your case, the possible dangers might not be worth it.

### HOMEOPATHICS AND FLOWER ESSENCES FOR EMOTIONAL HEALTH

If your new birds came from a scary situation, and especially if they had shock or trauma, one dose of the homeopathic

## The Adopted Chicken's Quarantine Protocol

◯ Add EVP to water daily for 2 weeks or 1 tablespoon of ACV per gallon of water daily.

◯ Add one clove of garlic and a sprinkle of oregano and thyme to one scrambled egg, and feed two or three times per week.

◯ Add two drops of Rescue Remedy or other applicable flower essence in water daily (can be used with EVP or ACV).

◯ Treat topically for external parasites.

◯ Supply new chickens with dust baths, and make sure the established flock has a dust-bathing area.

aconite in a 30c potency may help (see Chapter 9 for how to administer homeopathics). You can also give your new chickens some Bach Rescue Remedy or similar flower essence combination when they arrive and for a week or two after. For specific traumatic situations, the following single flower essences may help.

**Rock rose** is for chickens who are extremely afraid and panicked. They act as if they are about to be killed and might even harm themselves trying to escape their enclosure. They will work themselves into a frenzy and then appear completely exhausted.

**Aspen** is for chickens who are anxious and shy and lack confidence. They may seem to want to integrate but are too afraid. This is good for single chickens being integrated into a new flock or for chickens who lost their flock and therefore must be integrated into a new one.

**Star of Bethlehem** is for chickens who seem to have PTSD from trauma and grief, either recent or long ago. It seems they just can't let go of the incident—like they are stuck in the traumatic moment. This is a good remedy for chickens who may have been abused or neglected.

**Vine** is for chickens who relentlessly bully new additions to the point of not letting them eat or drink even after the integration process. They may force new chickens into constant hiding. These chickens are too dominant, going overboard with the pecking order.

**Chicory** is for rude roosters who are possessive and overbearing with new additions (or existing hens). They may refuse to let a new hen integrate or try to mate her constantly.

## The Process of Integration

It takes some strategy and a bit of patience to add new baby chicks to a flock if they weren't hatched by a mother hen who is already an established flock member. The younger a chicken is, the more submissive it will be, making it more vulnerable to domination and bullying from older hens and roosters. For this reason, wait until new chicks are fully feathered and either almost the same size as the established flock members or of adult age before you even think of beginning the integration process.

For instance, if you are adding standard chicks to a flock of standard chickens, the newbies will be ready to begin integration at about 12 weeks. Even though the new chicks are young and likely still docile, they will be large enough to avoid being completely oppressed by the older chickens. However, if you are adding tiny bantams to a flock with standard-size birds, you will want to wait until the bantams are 16 to 20 weeks old. As older hens, they will be less willing to put up with domination and bullying regardless of their size, making integration easier.

The process for integrating new chickens begins with placing the new members in a nearby area where they can see the established flock and the flock can see them, but they can't touch each other. It's best if the two groups can be as close as possible to each other. You want them to be able to go through all the pecking order rituals like posturing and lunging before they can physically touch each other. I call this having them "separate but seeing." You can use fencing or wiring to separate them; just make sure they can't breach it.

Keep this setup until your established chickens seem to be ignoring the new ones. If you have a very small flock with a gentle hierarchy, this process may only take a week. If your flock is bigger and very diverse, it could take up to a month. Other things that will affect the timing are the age, size, and temperament of the newbies. The key is to be patient and watch their behavior closely.

Some folks like to add their new chickens to the roost at night, after the established flock is sleeping. I prefer to open the partition right before dusk, allowing the new chickens to find their place on the roost. It is important to watch them during this time; you should see a bit of

## When Can I Integrate My New Chicks?

Note that you may be able to integrate your chicks sooner than outlined below, or you may need to wait longer, depending on the unique personality of your flock.

**Standard to standard:** At least 12 weeks

**Bantam to standard:** At least 16 weeks

**Standard to bantam:** At least 10 weeks

**Adult to adult:** Any age

squabbling, but since everyone will be working on going up to roost, the quarrels shouldn't be too bad. Return to the coop at sunrise to observe your newly integrated flock's behavior, and make sure the new members aren't being bullied in a drastic way.

It's sometimes tough for new chicken keepers to differentiate between actual bullying and normal pecking order behavior. Even with a long "separate but seeing" process, the pecking order still needs to be worked out. This means that chickens will squabble and chase each other, and more docile or young chickens might hide. Make sure there are plenty of places where new chickens can retreat from inhospitable birds. Provide more than one food and water station so that all those involved can eat and drink, and be sure to offer lots of treats to distract them.

Many people who love chickens have a really hard time witnessing this process, but rest assured, as long as they aren't drawing blood on each other, they are doing exactly what they need to do to reach peace again. If they are fighting so much as to draw blood, you will need to separate them and start the integration process over again. But don't worry! They will get there.

## Hatching or Adopting Baby Chicks Naturally

If you want to have new baby chicks raised within your flock and you are committed to avoiding overcrowding and have a plan for unwanted roosters, I believe the best method is to allow one of your hens to hatch out a clutch of fertile eggs or place day-old baby chicks under her to adopt. Chicks raised by a mother hen are generally healthier and more resilient than those raised in an incubator, plus they bring delight to humans. Watching a mother hen raise baby chicks is one of the most rewarding experiences of chicken keeping. For children and adults alike, it's a wonderful lesson in nurturing, dedication, and the cycle of life.

### HATCHING OUT

In order for a hen to hatch out eggs and raise babies, she must first be broody, meaning her instinct to sit on eggs and brood babies has been activated. You will know a chicken is broody because she will stay in the nesting box all day long, make a terrifying noise if you try to take eggs from beneath her, and possibly peck you in the process. Her belly will be warmer than normal, and she may have some bare spots on her breast and tummy where she plucked her own feathers in preparation for skin-to-egg contact.

If you catch a chicken going broody and you want her to hatch out chicks, it's time to locate some fertile eggs—unless you have a rooster, in which case you likely already have some. You can usually find fertile eggs through social media chicken groups or online classifieds.

Watching a mother hen lovingly raise her chicks is one of the great joys of chicken keeping.

Once you acquire fertile eggs, it's wise to date each one with a pencil (do not use a pen or marker, as the ink can be absorbed through the shell) so that you know which ones are for hatching and which may have been recently laid by another hen. Place each fertile egg in the nesting box next to your broody hen. She should gratefully roll them beneath her, beginning the approximately 21-day process of incubation.

If the brooding hen is young, observe her a few times a day to be sure that all is going smoothly. Sometimes young and inexperienced hens will abandon a clutch of eggs, allow them to get chilled by staying off the nest for too long, or even break and possibly eat the eggs she was meant to mother. You might become concerned that she is not eating or drinking, but as long as she has access to food and water, rest assured that she is fine. She's just waiting until you are not around to grab a bite to eat and take a quick sip of water. She doesn't want to risk you or any other human stealing her eggs (you really can't blame her since you've done it before)!

One of the advantages to having a hen brood babies is that she will take full responsibility for them, doing everything she can to keep them safe both before and after hatching. That said, it's important that she be in a predator-proof area that isn't so high off the ground that chicks could become injured after they hatch. Most mother hens will fiercely protect their young, and most flocks understand that baby chicks are off-limits when it

That first crack in the egg is called the pip. Soon your baby chick will enter the world!

comes to their pecking order shenanigans. But if your broody hen is low on the pecking order or you have unruly bullies or cruel roosters, you might want to keep Mama and babies in a safe, separate area (preferably where they can still see and be near the flock) until you feel comfortable that the biddies will not be harmed by their elder flock members.

On day 20 or 21, the chicks should begin to pip (make a hole) through their shell. It usually takes a couple of days for all the chicks in a clutch to hatch. This is the initial phase of hatching and is certainly an exciting time, especially for young children who have been patiently waiting for this event. But remember: Hatching is hard, stressful work. It can take about 24 hours for a chick to make it out of their shell. While it's okay to check on the mama, take some photos, and pick up the chicks occasionally after hatch, remember to honor their process. Give the mama and her chicks privacy most of the time so that they can concentrate on the world ahead of them.

Remember baby chicks can live without food and water for a couple of days after hatching, so don't worry if the babies choose to stay beneath Mama rather than venture to a nearby feeding and watering station. When the mother hen knows that everyone who is going to hatch has done so, she will move her babies out of the nest and into the world in search of food and water. In my opinion, this is where the fun really begins. I love watching a mama hen with her babies walk about the yard together.

## ADOPTING CHICKS

If you don't want to wait for a hen to hatch out eggs and/or you'd rather lower your chances of unwanted roosters, you can acquire sexed chicks and place them under a broody hen. To do this, you must be sure to get very young chicks; for the best chance of success, they should be younger than 3 days old.

Keep the chicks under supplemental heat until the sun has gone down, then place them in a box they can't jump out of, without heat, next to your broody hen. They should be close enough for her to hear them. The chicks should start to make a distressed call, which should prompt your broody hen to respond to them with a cooing sound. This is her way of calling them over to snuggle beneath her and be mothered. It's also your cue to place the babies beneath her. Once they are tucked under her, watch them closely

---

### How to Get a Hen to Adopt Baby Chicks

First, start with chicks that are younger than 3 days old.

1. At night, place the chicks in a box next to your broody hen.

2. Wait for the hen to make cooing sounds in response to the distress call of the chicks.

3. Tuck the babies beneath the mama.

4. Check to be sure she is not harming them.

5. Observe them at first light to be sure all is well.

to make sure she is not attacking them (this can happen, especially with inexperienced mothers). You also want to make sure to check in on them right at dawn. If all is well, she will be caring for them as if they were her own.

If you find that she has rejected any chicks or, heaven forbid, that she has harmed any of them, quickly bring them inside and get them back under artificial heat. Due to the fact that a small percentage of hens reject their chicks, it's important to have a brooder ready and be prepared for the possibility of raising the biddies yourself.

### RAISING THE NEW CHICKS

Both baby chicks and Mama Hen can eat starter feed, and in fact, you can switch the whole flock to starter during this time if everyone is integrated together. Continue to provide oyster shell or another calcium supplement for your laying hens, preferably in an area where your baby chicks will not access it, although they likely will not eat it even if they could. If your chickens are living in an enclosure where there isn't a lot of sand or tiny pebbles, provide chick grit by sprinkling some in their food and around the run. Mama Hen will make sure they eat it.

When your baby chicks are around 6 to 8 weeks old, you may observe the mother hen rejecting them somewhat. This is because she is preparing to lay soon and no longer wants the burden of caring

for younglings, especially since they have become too big to snuggle under her and all she wants to do is spend some kid-free time on the roost (so relatable!). At this point, other flock members may scold the young chickens, teaching them their place in the pecking order. It should not be too severe, though. Generally, it's much easier for young chickens to integrate into a flock when they've been raised there as chicks. In the unlikely event that the older chickens are displaying problematic behavior, such as relentlessly bullying or causing injury to the new additions, go ahead and begin the process of integration as discussed on page 158.

If you have a rooster in your flock and one or more of your baby chicks is a rooster, the established rooster *should* accept them as long as they know their place, but this may vary by breed and individual personality. It's also possible that the roosters may fight once the young rooster's hormones begin to kick in, usually around 6 months. This is another reason why it's important to have a plan for roosters. Even if you have the space and the right hen-to-rooster ratio, it's possible you may need to rehome someone or move them into a separate area once they reach maturity.

## The Rewards of Rescue

While I prefer to rescue chickens, I'm not against purchasing chicks through a breeder or buying them from a farm store.

I still do it on occasion, and I know how convenient and fun it is. But in the back of my mind, I know that chicks sold at farm stores and through breeders don't always go to so-called forever homes. Also, many well-intentioned hobby breeders hatch out too many chicks, only to scramble to place them in homes after the fact. There are just so many unwanted chickens out there! While it wouldn't make sense for me to tell people to *only* rescue chicks and never purchase them or hatch them out, I do think it's important to spread the notion that giving homeless chickens a better chance at life is a rewarding experience both for the rescued and the rescuer.

If you feel that you have the physical and emotional capacity to adopt rescue chickens, it's important to do so with the full knowledge of what it could involve. As I've already noted, there is an increased chance that you will introduce disease to your flock. Anyone who decides to rescue must fully understand this risk. Some rescue chickens may have also experienced trauma, not been socialized at all to humans, or lived their entire life in a tiny cage. If this is the case with your rescue chickens, you will need a certain amount of determination to overcome these issues. You will also need to have a steadfast belief that even the most mentally broken chickens can be brought into a balanced state with a little faith, patience, and knowledge of what to do. If you're not ready to take on such a challenge, that's 100 percent understandable and okay. If you are ready, I think that's wonderful and I wish the very best of luck to you and your flock.

# THE SEASONS OF LIFE

## and Chicken Keeping

One of the most valuable lessons in chicken keeping, especially for young children, is that life is full of seasons. Of course, there are the actual seasons of spring, summer, winter, and fall, each of which requires special considerations when raising chickens. But a flock also reflects to us the seasons of life: birth, youth, old age, and death. Having chickens is a constant reminder that while the joys of life are fleeting, they always return. Even the seemingly simple cycle of the egg (it's actually rather complex) represents a season of beginnings and ends.

## Eggs and the Cycle of Life

I cannot tell you how accomplished I felt when I peered into the nesting box and gazed upon that perfect miracle of my chickens' first egg. Yes, I took pictures, and yes, I posted them all over social media. I mean, I know I didn't lay the egg, but I had put so much care and time into my chickens by that point that I almost forgot about the whole reason I originally sought to acquire them: breakfast. The excitement of holding your first egg is something you either already understand or will understand soon enough. And when you eat it, well, that will be the beautiful bow that ties this whole chicken thing together. Trust me.

While chickens on average begin laying anywhere from 16 to 20 weeks, birds bred for show or other qualities can sometimes take longer to produce that first egg. The time of year you get your baby chicks and when they come into lay are also factors. For instance, if you bring your chicks home in February, you will likely get your first eggs in the late spring or early summer. But if you bring them home in July, they will reach laying age during a time when the days are shorter and chickens, in general, are not laying. In that case, you may not see your first eggs until spring.

When young hens are about to start laying, you may observe their combs and wattles getting redder and more pronounced. When you approach them, they may squat in a submissive position (they think you are a rooster and want to experience some fertilization—awkward, I know). Young hens coming into lay soon may also get rather noisy, pacing near the nesting boxes like they know they need to do something but aren't sure exactly what that something is. They may sing the egg song, even though there is no egg to celebrate yet. Think of it as the "pre-game show."

When hens do finally begin laying, don't be alarmed if you observe

There's nothing quite like discovering that first egg in the nesting box!

abnormalities such as shell-less eggs, which are often called soft eggs, or very small eggs, which can be referred to as fairy eggs. A young hen may lay a few days in a row and then suddenly stop for several days, causing her humans to worry that she has a health issue. As long as your chickens are acting normally and are eating and drinking, try not to worry. Laying eggs is tough work, and some chickens just need a little time to get into the groove of things.

## To Light or Not to Light

During the winter months, your chickens will likely take a break from laying. You might think this pause is due to the drop in temperature, but it's actually due to the lower number of daylight hours your chickens experience during this time. Unless she is an unusually prolific layer, a hen normally needs at least 14 hours of daylight during a 24-hour period to trigger her ovulation cycle. Without this amount of light, her body will instinctively stop the laying process and wait for the longer days of spring. This hormonal function happens to be a very handy feature for the species as a whole. Since it's more beneficial for chicks to hatch out during a time of year when the weather is mild and the foraging plentiful, chickens have enjoyed a high chance of survival throughout the generations.

Because I feel that a chicken's natural cycles are there for a reason, I allow them to have their winter break from laying. Some chicken keepers, however, decide to give their chickens artificial light during the winter months so that they will continue to lay. Chickens that are pushed to lay have a higher chance of developing ovarian cancer, and the presence of certain types of lights can cause stress-induced issues such as cannibalism, hyperactivity, and aggression. All that said, if you can't afford to lose that protein source during the winter months, or if you have an egg business, I encourage you to offer supplemental light with a clear conscience. Remember: The health of our chickens is connected with the health of our family and of our community. Weighing risks means taking all of that into account.

If you decide to supplement light, there's an optimal way to do it safely and with the least amount of stress on your flock. First, consider your birds' natural rhythm. If you were to watch your chickens as the sun goes down for the day, you would see them slowly moving toward their coop, pecking and scratching along the way, perhaps having a quick bite to eat and a bit to drink, then spending several minutes looking up at the roost, pondering where they will sleep for the night. They may pace about, chatter loudly, or squabble over who will retire where. The whole process is slow and seems calculated, because it is.

It's important for your chickens to have a chance to find their place to roost and exercise their natural pecking order instincts, both of which occur at dusk, so I recommend supplementing light only in the morning hours. It's also pretty distressing for a chicken when the lights are suddenly turned off and they haven't settled to roost.

If your goal is to have eggs through autumn and winter, it's best to begin supplementing light in the fall, when the days are just starting to get shorter. Once daytime shrinks to 14 hours, add lighting in the morning as needed to keep that period from getting shorter. Alternatively, if you decide to give your chickens a natural fall, you can begin supplementing light after the winter solstice. In this case, add your light in 30- to 45-minute increments weekly until you get to 14 hours. The easiest way to keep your light length consistent from day to day is to use a timer, though you will still need to adjust it with the changing day lengths both going into shorter days and coming out of them when the need for supplemental light decreases.

A "warm" LED light (as opposed to "cool") is better for egg-laying; this could be a single bulb, Christmas lights, or a nightlight. Place the lighting above where the chickens roost, either in the center of the ceiling if using a single bulb or along the edge of the ceiling if using string lights, and be mindful that there is enough light for your chickens to operate as they would

### The Dos and Don'ts of Supplementing Light

**Do:** Allow a natural dusk.

**Do:** Use warm LED light.

**Do:** Gradually work up to 14 hours if the days are already shorter.

**Don't:** Use a heat lamp as supplemental light.

**Don't:** Add light in the evening.

**Don't:** Leave the light on all the time.

during the day. *Always* make sure that whatever lighting you use is far enough away from shavings, straw, or feathers to avoid a fire hazard. You will also want to wipe dust and dirt off them weekly.

## Molting

Few things send a panicked shudder down a new chicken keeper's spine like suddenly finding feathers scattered all over their chicken yard. Naturally, their first thought is that something awful has happened to one of their birds. But what if, after counting their brood, they find that all members of the flock are present and happily pecking and scratching? Depending on what time of year it is, it's possible that their chickens have begun their annual molt.

When chickens molt, they drop their worn-out feathers and replace them with shiny new ones. This usually happens in the late summer or early fall, so the chickens will have better protection from the cold in the winter months, but it has been known to sometimes occur during other times of the year and even after periods of intense stress or illness. Also, there always seems to be one wacky chicken that molts in the dead of winter (usually a Frizzle, bless their hearts).

When chickens are babies, they're like furry little cotton balls. Gradually, their fuzz is replaced with feathers during several molts as they grow into adulthood. During their awkward teenage phase (about 8 to 12 months), they can look especially rough. If they are getting the proper nutrition and have enough space and their stress levels are low, they should blossom into full-feathered glory in no time.

It normally takes chickens about a month or two to get through their molt. Most hens do not lay during this time because so many of their bodily resources are going toward building new feathers for the winter months ahead. If you observe a molting chicken that seems lethargic or sickly and you've ruled out other issues, bring them inside and administer the R.E.S.T. method for a few days until they recover their strength. To prevent health issues related to molting, you can

Molting chickens can appear rather unsightly! Rest assured it's a natural and necessary process.

supplement a little extra protein and give them a 2-week course of EVP or your homemade electrolytes. For example, you can go from a 16 percent feed to an 18 percent feed, or you can just offer some healthy protein treats like scrambled eggs, grubs, or mealworms. Be mindful not to

Young chickens go through several molts before reaching adulthood.

overdo the protein, though, because it can cause other problems such as digestive issues.

Of course, molting is not the only cause of feather loss. Other causes include bullying, overactive roosters, external parasites, or feather picking. Make sure that you rule out these issues if you suspect they may be at play. Since molting can be stressful to chickens and stressed-out chickens are more susceptible to parasites, I recommend regularly checking your chickens' bodies for parasites during this season. You may also want to be extra diligent about implementing your immune-boosting practices during this time and be especially mindful that they have adequate dust-bathing areas to ward off freeloading parasites.

## The Challenges of Winter and Summer

The seasons that require the most mindful care are winter and summer. In the winter, there's the question of how low temperatures can go before you have to intervene. In the summer, there is the legitimate fear of overheating. Even if you've chosen the appropriate breeds for your climate, there can be times when you need to step in and protect your flock from harsh weather. Let's begin with winter.

### WINTER
Moisture in the winter is more dangerous for your chickens than low temperatures.

If there isn't good airflow in your coop, condensation can develop not only from the dampness in the air but also from your chickens' breath and droppings. Moisture, along with an increased chance of ammonia buildup, can lead to respiratory as well as other health issues. Condensation and low temperatures are also a recipe for frostbitten combs, wattles, toes, feet, and even legs. This is why I recommend making sure your coop has excellent ventilation, even in the winter when opening vents feels counterintuitive. I also recommend keeping waterers out of your coop when possible, as they can raise the moisture level of the air.

During the colder months, you can give your chickens a treat with carbohydrates, fat, and protein before bed to keep them warm through the night, since the digestive process produces heat. Scratch grains or cracked corn are great, and if you can mix in a little fat, like grubs, that's even better. You can also skip purchased treats and instead focus on healthy scraps with a little more carbohydrate content, like your family's leftover brown rice or pasta mixed with a small amount of a healthy protein. Just make sure your chickens eat any scraps before roosting time because

A healthy snack before bedtime can help your flock stay warm through the winter nights.

rodents are especially desperate in the winter. A tasty treat may lure them into the coop, to discover more tasty treats. And be mindful of the Chicken Food Pyramid guidelines (see page 104) so that

your chickens' diet doesn't get out of balance (but don't get too obsessed with following the pyramid to a tee, because that's not balanced either).

The deep litter method is a wonderful way to keep your chickens cozy when it's cold out. The composting process generates heat, and you will be surprised by the temperature difference in a coop with deep litter versus one without. When done correctly, deep litter also makes for a healthier coop environment in general; the beneficial bacteria it encourages ward off common cold-weather issues such as ammonia buildup and pathogens. The deep litter method is simple to learn and execute, but if you have any issues, it's best to start the process over or cease practicing it until you can get the hang of it. When deep litter isn't functioning correctly, it can be harmful to your flock, especially in winter.

If your coop has good ventilation and there aren't any drafts where your chickens roost, it's unlikely that your chickens will need supplemental heat, even when the temperatures get below freezing. Remember, chickens are wearing thick down coats and will huddle together on the roost to stay warm. If you start to worry that your chickens are too cold, turn your thoughts to all the flocks in Minnesota, Manitoba, or Alaska that are getting through the winter without supplemental heat! All this said, there are exceptions to this rule.

## When to Supplement Heat

❖ You have elderly chickens or very young chickens.

❖ You have exotic or other non-cold-hardy breeds.

❖ You have chickens that are sick or have special needs.

❖ The temperatures have gone below –15°F (–26°C) for an extended period.

❖ You have done everything correctly, but due to environmental or other issues, you cannot control the moisture in your coop.

If you need to supplement heat, do so safely with a radiant-type heater made specifically for chicken coops. You can find heaters for almost any budget, and you can possibly find them used if you're willing to spend some time searching.

## SUMMER

Much emphasis is placed on getting chickens through the winter safely, but high temperatures can actually be more dangerous for chickens than low ones. Overheating and heat stroke are unfortunately a common reality during the hotter months, so pay special attention to keeping your chickens cool, even if they are heat-hardy.

Summer is yet another time when ventilation is of the utmost importance. Aside from the normal ventilation provided in your coop, you may also want to open windows and doors during the day, and, if possible, at night. Make sure your coop is still predator-proof by covering any ventilation openings and windows with hard-wire mesh (or, at least, doubled-up chicken wire), and only keep the coop door open if your run is completely predator-proof.

Hydration, of course, is key; if there is ever a time to make sure your chickens have clean, cool water, it's during the summer months. Take care that your watering station is not in direct sunlight, and change it as often as needed. You may also wish to add additional stations to ensure all members of your flock are able to drink as much as they need. During heat waves, stop giving your chickens ACV (it could affect their calcium levels in hot temperatures), and instead supplement with a week or two of EVP or your homemade electrolytes. Don't keep them on electrolytes for the whole summer, though, as extended periods of consuming supplemental electrolytes can create an imbalance. Fermenting their feed is another excellent way to keep chickens hydrated during this time.

Plant chicken-resistant trees and bushes or build your run around existing shade; this will work wonders when the sun is at its hottest. You can also invest in some tarps (you can often find them for free) and place them around the chicken yard to create extra shady spots. Placing sand under the tarps encourages your flock to dust bathe in a cool spot.

## HEAT WAVE PRECAUTIONS

If you are experiencing a heat wave, take the precautions listed on the facing page to offer your chickens some relief. When and how often you need to take these steps depends on the breeds you have and the climate you live in.

For instance, where I live in Northwest Washington State, the seasons change gradually, with few surprises. This gives my chickens a chance to acclimate slowly to the summer temperatures, which usually peak around 90° to 95°F (32° to 35°C). I rarely have to do more than make sure my chickens are well hydrated and have lots of shade. Occasionally, I leave misters on for them and offer them some frozen fruit or cold greens in their water bowl.

Several of my breeds aren't considered heat-hardy and they do just fine.

My sister, however, lives in North Texas, where temperatures can go from 60° to 90°F (16° to 32°C) in a matter of hours. They can peak in the 100s with little cool-down even overnight. The climate where my sister lives is far more dangerous for chickens, especially for heavier breeds with feathered feet and pea combs. We both have to take precautions, but my sister must be extra diligent. Not only does she often have to follow many steps in the heat wave action plan, but she must keep an eye on her chickens periodically through the hot days to be sure she doesn't lose members of her flock.

## Heat Wave Action Plan

✤ Offer EVP or homemade electrolytes in the waterers.

✤ Add ice or frozen berries to waterers throughout the day.

✤ Use misters or spray down the run several times a day if your climate is dry.

✤ Offer shallow pools of water for chickens to stand in.

✤ Freeze gallon jugs of water and place them in the coop or other shaded areas.

✤ Offer hydrating fruits and vegetables such as cucumber and watermelon.

✤ Offer hydrating greens such as lettuces straight from the fridge.

When chickens are hot, they may pant, flutter their throat muscles, or hold their wings out to the side in an effort to keep their bodies from overheating. Some panting and wing-fanning are normal during hot temperatures, but if your chicken is constantly displaying these behaviors, is listless, or appears disoriented, you must take immediate action to avoid overheating or even death.

## Heat Stress Action Plan

1. Bring the affected chicken inside where it is cool.

2. Place the chicken on a cool, damp towel.

3. Pat their comb, wattles, and feet with a cool, damp towel.

4. Offer them electrolyte water, or if they're not drinking, pool a few drops of electrolyte water onto the side of their beak via a needleless syringe.

5. Give your chicken three doses of the homeopathic remedy aconite in 30c potency (optional; see page 147 for how to dose homeopathics).

## The Final Season: End-of-Life Care

All seasons must come to an end, and all chickens must eventually peck and scratch their way to the great coop in the sky (or wherever you believe their little spirits go). I've been doing this poultry thing for a long time, and no matter how often someone tells me they are "just chickens," I still grieve and feel the shock of losing a beloved pet when one dies. When it happens, I am also reminded of how much the Western world struggles to face the reality

that our bodies won't be here forever. It's helpful for me to remember that dying is as natural a process as any other function we experience on this earth. When it comes to end-of-life care, I've stopped desperately trying to prevent the inevitable and have instead shifted my focus to making their transition as peaceful and as smooth as possible.

Any chicken keeper *must* have a plan for the end of a flock member's life. If one of your birds is suffering and needs to be euthanized, do you have someone who can help you? Do you have access to and savings (if possible) for veterinary care? If you're the type who can perform this task yourself, have you researched the best methods? At the very least, it will be a good idea for you to designate an area in your home, garage, shed, or coop where a sick or dying chicken can be comfortable as they transition to the next phase. You may want to have on hand some homeopathic remedies and flower essences that can help keep your chicken calm and comfortable through the process.

This may come as a surprise to you, but I can't euthanize my own chickens. I don't often admit this, because when I have, I've been met with harsh judgment and even told I shouldn't own chickens if I'm unable to fulfill this important task. In my defense, and in yours as well if you find yourself in this position, not all people come to this earth with the same skills. I've known for a long time that my role in this world doesn't include ending a pet's life, even when it's the humane thing to do. This is not because I think it's wrong; on the contrary, I find it admirable when someone is able to put an animal out of its misery. It's just not something I can do. This is who I am, and I believe that's okay.

Whenever someone tells me that they lost a flock member, my response is always the same: That bird's life, however long or short, was one less life spent languishing in a cage on a factory farm. I'm sure if birds could talk, your chicken would express how grateful they were for the experience of being cared for under such humane circumstances and by such a compassionate human. I hope when you are faced with loss in your flock that you can take these sentiments to heart.

---

## Homeopathics and Flower Essences for Ease in Dying

❖ **Angel's trumpet.** This flower essence helps a chicken surrender to and have strength through the process of dying. Rub two drops on their back several times a day.

❖ **Bach Rescue Remedy or other similar flower essence.** Use to help relieve general anxiety. Rub two drops on their back periodically during the process.

❖ **Arsenicum album (30c potency).** Use for chickens that seem to be resisting death or experiencing tension or fear during the process.

See page 147 for dosage instructions.

## A FINAL NOTE
# CHICKENS CAN CHANGE THE WORLD

What does it mean to change the world? When I was a kid, I would imagine myself making grand gestures, like delivering passionate speeches to large crowds from a looming platform. As I got older and settled down, part of me felt that I had failed because my dreams of changing the world didn't come true like I thought they were supposed to. Having grown up disconnected from my roots, I didn't understand that I had to rediscover my roots to fulfill the promise of my future. That is, until I held my first baby chick. At

that moment, to my utter surprise, a new dream took hold.

Through spending time with my flock, learning about my own history, and sharing what I've learned with whoever will listen, I now see the beautiful way in which everything is connected. Now I know that effecting change in the world can be as small as allowing one blade of grass to grow, raising one flock of chickens outside of a factory farm, or teaching one child how to gently gather eggs for the first time. I truly feel that everything I do in

Chickenlandia contributes to my dream of a better world, and I hope you can feel that, too, in your own chicken yard and beyond.

And you want to know something? I almost didn't write this book because I was afraid. I thought, What if I don't know enough? What if I'm not smart enough? I'm not a scientist, a veterinarian, or a farming industry expert. I'm just someone who really loves chickens. But I'm also human, which means there is knowledge in my history. It's the same knowledge that my great-grandparents Maria and Alberto used to survive the hardships of their time. And somewhere in your history there is a similar consciousness of resilience and survival.

Knowledge and intuition don't belong to an intellectual perched upon a hill. They aren't hidden in a room that only the privileged few can enter. They don't belong to corporations or any institutions. This ancient knowingness is yours. It's mine. It's *ours*. It's the wisdom of the earth and every time you enter your chicken yard, you gain more and more access to it. It's a gift to humanity that you have every right to welcome into your life.

No matter how chaotic and scary the world around you gets, no matter how divided humanity becomes, remember to practice and share your knowledge. Pass it down to your kids. Share it with your community. Share it in the country, the suburbs, and especially the city. Exchange your chicken ideas with those who don't agree with you on other matters. If you can focus on this one unifying thing, maybe other differences will become less of a roadblock. Maybe for you and the lives you touch, those differences can be replaced with an understanding of our own stories and consequently what it means to be human. This understanding is the key to a better future for all of us.

And maybe, by caring for our flocks and in turn our families, our communities, and our planet, you and I really can change the world.

# RECIPES FROM MY MOTHER

I'm pleased to share with you these authentic Guatemalan recipes, which come from my mother, Dicla, who so generously shared them with me (though they never seem to taste the same as when she makes them). I grew up eating these dishes while listening to the music of Chapinlandia, and they are even more delicious with fresh eggs from my flock. I hope you find them as yummy as I do. My mother's salsa picante recipe is especially famous among our family and friends!

## Ejotes Envueltos con Salsa
### (Folded Green Beans with Sauce)

This recipe is a staple in our family. Whether it's Thanksgiving, Christmas, or someone's birthday, you could find it served as a side dish in our home. The most popular way to prepare it is with green beans, but I've had other vegetables "folded con salsa" as well, including one time when my grandmother served us folded watercress. Once you learn this recipe, you'll find you can try it in other ways and really make it your own. Enjoy!

### FOR THE SAUCE

- 4 ripe Roma tomatoes
- ¼ medium yellow onion
- ½ red bell pepper
- 2 cups water plus more as needed
  Salt and pepper
- 1 teaspoon olive oil

### FOR THE GREEN BEANS

- 1 quart water
- 1 tablespoon salt
- 1 pound fresh green beans
- 2 eggs, whites and yolks separated
- 2 tablespoons avocado oil

1. To make the sauce, place the tomatoes, onion, and bell pepper in a saucepan and cover them with the water (add more water if needed to cover them). Turn the stove to medium-high heat and bring the water to a boil. Add salt and pepper to taste. Boil until the tomatoes, peppers, and onion become soft, then set the pan aside (do not drain) to cool down.

2. Using a slotted spoon, remove the cooked and cooled tomatoes from the pan, peel, and add them to a blender. Add the cooked onion, pepper, and cooking water. Blend on high until smooth.

3. Heat the olive oil in the same pan used for the tomato mixture. Add the blended vegetables to the pan, and boil until thickened, stirring frequently. Add salt and pepper to taste.

4. To make the green beans, place the water and salt in a medium saucepan and bring to a boil. Add the green beans, and cook until just tender, about 5 minutes. Do not overcook. Drain, then set aside on a plate or cutting board to cool down.

5. Separate the cooled green beans into small bundles, about six to eight green beans each. Place your bundles on a clean kitchen towel until they are completely dry.

6. Meanwhile, beat the egg whites with a handheld mixer until they form firm peaks, then slowly add the yolks and beat until the mixture is uniformly yellow.

7. Heat the avocado oil in a frying pan on medium-high heat until hot but not smoking. Fold each bundle in the egg batter to coat completely, then place them in the oil, turning them periodically until all sides are golden brown, about 3 minutes. Do not overcook.

8. Set the fried green beans on paper towels or a clean kitchen towel to drain the excess oil. Serve warm with warm sauce on the side or poured over them.

## Huevos Revueltos con Cebolla y Tomate
### (Scrambled Eggs with Onion and Tomato)

Growing up in a Guatemalan household meant that the conversations must be lively, the music must be rich, and the food must be as colorful as possible. Even the scrambled eggs had to have a little pizzazz. This simple recipe is beautiful and chock-full of flavor. Add some raw chopped cilantro and yellow onion if so desired, and while you're at it, you can serve this with a nice mug of Guatemalan coffee!

    2  teaspoons butter
    ¼  cup chopped yellow onion
    4  Roma tomatoes, diced
    8  eggs
       Salt and pepper
       Corn tortillas (optional, for serving)

1. Melt the butter in a frying pan over medium-low heat. Add the onions, and sauté until translucent.

2. Reduce the heat to low, and add the tomatoes. Cook until the tomatoes are soft, stirring often, about 2 minutes.

3. Crack eggs into a bowl. Pour the eggs into the pan and stir until cooked but still soft, about 2 to 3 minutes. Add salt and pepper to taste. Serve hot with warm corn tortillas on the side.

## Dicla's Salsa Picante

I confess that I was a tad nervous asking my mother for this recipe because I thought for sure that she would want to keep it a special family secret. Of course, my mom, being the no-nonsense woman she is, simply laughed when I asked her about it and told me to share away! So here it is. This tasty salsa is especially delicious on fried eggs, but you can add it to any dish that you think would be good with a little salsa picante. Just do me this favor: When your friends compliment you on this recipe, tell them it's from my mom.

    1  teaspoon olive oil
    ¼  cup chopped onion
    1  jalapeño pepper, chopped, with seeds removed
    1  (13.5 oz) can diced tomatoes or 1¾ cups peeled, fresh tomatoes (see Note)
    2  large or 4 small tomatillos, quartered
    ½  teaspoon salt
    ½  cup chopped fresh cilantro

1. Heat the olive oil in a frying pan over medium-low heat. Add half (⅛ cup) of the onion and the chopped jalapeño to the pan. Sauté until the onion is translucent, about 2 to 3 minutes.

2. Add the tomatoes, tomatillos, and salt. Lower the heat to low, then simmer for about 10 minutes. Remove the pan from the heat, and cover. Let cool.

3. Add the cooled mixture to a blender and blend on low until just combined (you want it to be a tad chunky, not smooth). Place the mixture in a bowl, and add the remaining raw onion and the chopped cilantro. Stir to combine.

Note: If you are using fresh tomatoes, place them in boiling water for 30 seconds, then remove, let cool, and peel. Cut them in half before blending.

## INTERIOR PHOTOGRAPHY CREDITS

# INDEX

Page numbers in *italics* indicate illustrations and photographs.

# EMBRACE THE CHICKEN-KEEPING EXPERIENCE

## with More Books from Storey

### How to Speak Chicken
#### by Melissa Caughey

*Do chickens have names for each other? How do their eyes work? How do they learn?* This quirky and fascinating guide is full of insights into how chickens communicate, understand the world, establish roles within the flock, and more.

### My Chicken Family
#### by Melissa Caughey

This unique keepsake album invites you to create a lasting record of your chickens' lives and adventures, with write-in prompts and spots for photos and stories of your flock as it grows. A perfect gift for any chicken lover!

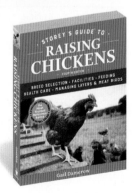

### Storey's Guide to Raising Chickens, 4th Edition
#### by Gail Damerow

For more than 20 years, poultry raisers have relied on this best-selling reference for comprehensive and expert information on shelter, breeds, food, health care, behavior, eggs, chicks, meat, and much more.

**JOIN THE CONVERSATION.** Share your experience with this book, learn more about Storey Publishing's authors, and read original essays and book excerpts at storey.com. Look for our books wherever quality books are sold or call 800-441-5700.